Cristofer Munguía
José Alvarez

Applied Geotechnics: Fundamentals, Methods and Design of Infrastructures

Cristofer Munguía
José Alvarez

Applied Geotechnics: Fundamentals, Methods and Design of Infrastructures

From ground exploration to foundation design

ScienciaScripts

Cover image: www.ingimage.com

This book is a translation from the original published under ISBN 978-620-9-36636-9.

Publisher:
Sciencia Scripts
is a trademark of
Dodo Books Indian Ocean Ltd. and OmniScriptum S.R.L publishing group

120 High Road, East Finchley, London, N2 9ED, United Kingdom
Str. Armeneasca 28/1, office 1, Chisinau MD-2012, Republic of Moldova, Europe
Managing Directors: Ieva Konstantinova, Victoria Ursu
info@omniscriptum.com

Printed at: see last page
ISBN: 978-620-9-73985-9

Contents

CHAPTER 1

Fundamentals of soil science

1.1 Definition and nature of soil

Soil is a heterogeneous natural material consisting of mineral particles, organic matter, water and air; it is the surface layer of the Earth's crust formed by geological and biological processes. In civil engineering, soil is considered 'a relatively loose aggregate of minerals, organic matter and sediments, deposited on bedrock'. Its study – *soil mechanics* – seeks to understand its behaviour, taking into account its physical and mechanical properties for the design of structures (foundations, embankments, retaining walls, etc.).

Soil can form through the weathering of rock (residual soils) or through the transport of particles (sedimentary soils) (see *residual soil*, *sedimentary soils*). Its thickness and properties vary according to climatic factors, bedrock and geological processes. In engineering, it is essential to anticipate changes in its properties over time (due to changes in pressure or water content), as soil is not inert but 'active and highly sensitive to environmental conditions'. A soil element is considered consolidated when it is in equilibrium under the maximum historical pressure that has compressed it (for example, due to overburden from upper strata); if the current pressure is lower, it is termed overconsolidated.

1.2 Soil classification and types

Soil classification is essential for describing their general characteristics in engineering terms. As there are countless types of soil, classification systems group soils into categories based on their properties (particle size, plasticity, etc.). These classifications provide a 'common language' for engineers when assessing the suitability of soil for different applications. Soils are commonly divided into two main groups according to grain size:

- Coarse (granular) soils: these include gravels and sands. Gravels

have grains larger than 2 mm; sands, between 2.0 mm and 0.05 mm. In engineering, the particle size distribution is analysed using sieves. Granular soils are considered non-plastic and, depending on their gradation, may be silty gravel, sandy gravel, well-graded or poorly graded, etc.

- Fine-grained soils: these include silts and clays, with particle diameters

less than 0.05 mm. Silts (0.05–0.002 mm) and clays (<0.002 mm) are fine-grained soils; clays exhibit colloidal properties and notable plasticity.

In addition to classification by particle size, there are specific engineering systems: for example, the Unified Soil Classification System (USCS) divides soils into coarse and fine fractions based on sieve analysis results and Atterberg limits (subdividing them into gravel (G), sand (S), silt (M), clay (C), as well as organic materials and peat). Another common system is the AASHTO (for roads), which groups soils into categories (A-1 to A-8) according to particle size and plasticity. Similarly, the USDA (agronomic) classifies soil texture (USDA Soil Texture) based on percentages of sand, silt and clay. In engineering practice, soil classification is usually presented in tables as: gravel (G), sand (S), silt (M), clay (C) or organic (O), with suffixes added to indicate well-graded (*W*) or poorly graded (*P*). The variability of soils means that there is no single definitive system.

1.3 Physical properties of soil

Physical properties describe the intrinsic characteristics of soil without considering external loads, and are critical for predicting its behaviour:

- Volumetric weight and density: The wet volumetric weight (apparent density) relates the total weight to the volume occupied by the soil. Dry bulk density (dry density) is calculated by removing the water. The specific weight of solids (grain density) is the ratio of the mass of solids to the volume they occupy. These properties determine the compactness and porosity of the soil. The void percentage (porosity) indicates the volume of voids relative to the total soil volume; its complementary measure is the void ratio (e), which is fundamental in soil mechanics.
- Natural moisture (water content): The soil contains water in its pores. Moisture content (the ratio of water mass to dry solids mass) significantly affects compaction and strength. Soils with high water content may have reduced strength (e.g. soft clays).
- Grain size distribution: Analyses the distribution of particle sizes. sieving and hydrometer tests are carried out (in accordance with ASTM/AASHTO standards) to obtain percentages of gravel, sand, silt and clay. From the particle size analysis, gradation curves are obtained which allow the soil to be classified (e.g. well-graded gravel, fine sand, sandy silt, etc.).
- Atterberg limits: For fine soils, the limits are determined liquid limit (LL) and plastic limit (LP), which characterise the plasticity of the soil. The plasticity index (PI = LL – LP) quantifies the extent of the plastic state. These parameters govern compressibility and plastic behaviour; for example, soils with a high LL are typically highly compressible and expansive.
- Permeability: This relates to the ease with which water flows through the soil. It depends on the size and connectivity of the pores. Granular soils are usually more permeable (sands, gravels) than fine-grained soils (silt, clays). This parameter is key to the design of drainage systems and the estimation of consolidation in saturated soils.
- Soil structure: Reflects how particles are grouped (flocculated, laminated, etc.). Clay soils may exhibit 'house of cards' or laminar structures, influencing their strength and volume.

1.4 Mechanical properties of soil

Mechanical properties describe the behaviour of soil under loads and deformations:

- Shear strength: This is the soil's ability to resist . In granular soils, internal friction (angle of friction φ) predominates; in cohesive soils, there is also cohesion (c) between particles. The typical formulation is the Mohr-Coulomb criterion: $\tau = c + \sigma \tan\varphi$. These properties are obtained through *direct shear* tests, *triaxial* tests, etc. Shear strength determines the load-bearing capacity and stability of slopes.
- Consolidation and compressibility: When a sustained load is applied to saturated soils (especially clays), gradual deformation occurs due to the expulsion of water from the pores. The process is described by Terzaghi's consolidation theory. The consolidation coefficient (Cv) and the compressibility index are measured in the consolidation test (edometer). Clay soils have high compressibility (significant settlement) compared to sands.

- Deformability (modulus of deformation): This is the soil's resistance to
stresses. It is expressed by the modulus of deformation (E) or shear modulus (G). Compacted or very stiff soils (well-graded gravels) will have high moduli, whilst soft soils (loams/loose clays) have low moduli. This parameter is used in settlement calculations and in the finite element method for soils.
- Bearing capacity: Represents the maximum stress that can
support the load without failure or excessive deformation. It is calculated using classical theories (e.g., Terzaghi, Meyerhof) based on c and φ, taking safety factors into account. A 'consolidated' soil supports loads better, whereas unconsolidated soils may experience delayed settlement.

In short, geotechnical engineering focuses on understanding these physical and mechanical properties of soil to ensure the stability of structures. In the words of experts, "soil mechanics deals with the study of the behaviour and the physical and mechanical properties of soils", and this knowledge enables the design of foundations, dams, embankments, roads and any structure that interacts with the ground. Through geotechnics, civil engineering focuses its efforts on these properties to "determine the behaviour and use [of the soil] in construction projects".

CHAPTER 2

Methods of analysis and applications of soil investigation

2.1 Geotechnical investigation and sampling

Before designing any structure, it is necessary to investigate the subsoil. This involves field surveys using boreholes and test pits to obtain representative soil samples. During borehole drilling, stratigraphy, water table levels and in-situ soil resistance are measured. Among the most commonly used penetration tests is the *SPT (Standard Penetration Test)*, where a sampler is inserted and the number of blows (N) required to penetrate to a certain depth is recorded; this value correlates with the density of sands and the strength of clays. Another common technique is the *CPT (Cone Penetration Test)*, which measures the tip resistance (qc) of a cone pushed into the ground and allows the soil to be classified and geotechnical parameters to be estimated.

In addition to traditional sampling methods, there are in-situ tests such as the wedge test (véleta) to measure shear strength in soft clays, or the pressuremeter (Ménard) to determine the modulus of deformation. The aim of these field tests is to characterise the actual (undisturbed) subsoil, identify critical layers (soft layers, expansive clay, etc.) and plan laboratory sampling. Samples may be undisturbed (more expensive, requiring tubes) or disturbed (sufficient for granulometric analysis). In any case, field data guides the design of safe and cost-effective structures.

2.2 Laboratory tests on soils

In the soil mechanics laboratory, numerous tests are carried out to quantify the properties of the soil. The main ones include:

- Grain size analysis: Sieving is carried out (for particles >0.075

mm) and the hydrometer method (for fine fractions) in accordance with standards (ASTM D422, AASHTO). The soil gradation curve is obtained, showing the percentages of gravel, sand, silt and clay. This is used to classify the soil (for example, as well-graded siliceous sand) and to calculate design parameters.

- Atterberg limits (plasticity tests): *The*

(LL) using a Casagrande apparatus and the *plastic limit (LP)* using the thread wrapping test. The plasticity index (PI = LI – PL) indicates the plasticity of the soil. These tests distinguish whether a fine-grained soil (clay/silt) is non-plastic, plastic or semi-solid, and are fundamental for classifications (e.g. determining whether a soil is SP, SC, ML or CL in the SUCS).

- Proctor test: Determines the *maximum dry density* and *the optimum*

for proper compaction. Samples are compacted with a defined energy in moulds at different moisture contents, and the resulting density is measured. In the design of embankment or pavement layers, these values are used to ensure stability (poorly compacted soil may suffer from settlement or erosion).

- Progressive Load Test (CBR): The *CBR (California Bearing Ratio)*

It is key in pavement engineering. It measures the resistance of a compacted soil sample to the passage of a piston: the CBR value is calculated by comparison with a standard material. A higher CBR value indicates soil capable of withstanding greater loads. This index is used to design the thicknesses of the subgrade and sub-base layers in roads.

- Shear strength tests:

or *Direct shear*: A prismatic sample is cut into a split box

, applying a constant vertical load and measuring the shear force. It allows the determination of c and φ of the soil under various confinement conditions. It is particularly used in relatively stiff granular and clayey soils.

o *Triaxial*: A cylindrical soil sample is subjected to confining pressure and

axial load. Depending on the drainage condition (UU, CU, CD), failure curves are obtained that determine c, φ and volumetric behaviour.

- Unconfined compression test (UCS): A cylindrical sample of

saturated soil is compressed axially without lateral confinement. The maximum pressure before failure is the unconfined strength σ_uσ. It is used in soft cohesive soils (clays, silts) as a quick indicator of strength; its typical use is to estimate preliminary bearing capacity.

- Edometric test (consolidation): A soil sample (usually

clay) is placed in a consolidation apparatus and subjected to incremental static loads, with vertical deformation measured over time. From this, the consolidation coefficient (Cv), modulus of compression and parameters enabling the calculation of deferred settlement under foundations are obtained. This test is crucial for clayey soils, as they 'exhibit high compressibility'.

- Special tests:

o *Los Angeles* and *Micro-Deval abrasion tests*: these assess the resistance to

of aggregates (gravel and sand) when rolling against each other (Los Angeles) or in the presence of water (Micro-Deval). Important for construction materials used in pavements and concrete.

o *Fines content (%) on No. 200 sieve*: Indicates the proportion of

fine particles (<0.075 mm) that affect compaction. Soils with a high fine content (above 10%) tend to have reduced permeability and require additives to optimise their performance.

These tests provide numerical parameters used in stability and design analyses. For example, the results of direct shear and triaxial tests provide the c-φ values for load-bearing capacity and slope stability calculations; the CBR test informs pavement design; and the Proctor test determines optimal compaction conditions to achieve the required maximum density. Taken together, "laboratory tests... are essential tools for the design and construction of safe infrastructure".

2.3 Applications in civil engineering projects

The knowledge gained from soil investigation is directly applied to the design and construction of structures. Among the most common applications are:

- Foundation design: The choice of foundation type (shallow

or deep) and their dimensions depend on the soil properties. Using c and φ measured in the laboratory and the bearing capacity method, the bearing capacity of the soil is calculated. For soft soils, piles or raft foundations may be required. Furthermore, based on the consolidation parameters and elastic modulus, the settlements (partial and total) of the foundation under permanent loads are estimated, thereby preventing excessive settlement.

- Slope and wall stabilisation: For embankment works or walls

For retaining structures, the soil's shear strength and active/passive earth pressure are analysed. The geometry of the slope and the geosynthetic reinforcements are designed taking into account the angle of internal friction. For gravity walls or gabions, the earth pressure is calculated using the same strength parameters.

- Paving and roads: In the construction of sub-bases and

sub-bases, the results of the CBR test or Proctor compaction test are used. A soil with low bearing capacity (low CBR) may require stabilisation (lime, cement, geotextiles) before being used in paving. Civil engineers use CBR values to determine the thickness of the granular layer in roads and airports.

- Deformation and settlement control: In sensitive projects

(tall buildings, bridges, dams) long-term settlement is assessed using consolidation theory. Pre-consolidation (pre-loading) or water drainage systems (vertical drains) can be designed to accelerate the process and minimise deferred settlement.

- Water, sanitation and environmental infrastructure: Soil permeability

soils is crucial in drainage, sewerage and landfill projects. Protective layers (bentonite, geosynthetics) are designed to prevent seepage in waste disposal sites.

In short, "civil engineers use applied soil mechanics to design foundations for buildings and bridges, to construct dams, and to stabilise embankments and slopes". Each project requires the integration of field and laboratory data to ensure stability and safety. The process culminates in a geotechnical report, which compiles the soil characterisation and provides recommendations (type of foundation, necessary treatments, design parameters).

2.4 Soil investigation report and recommendations

The soil investigation is summarised in a technical report that includes: a stratigraphic description; test results (grain size distribution, Atterberg limits, shear tests, consolidation tests, etc.); tables of geotechnical properties (unit weight, cohesion, angle φ, modulus of compression, etc.); and conclusions. This report is essential for civil engineers, as it "provides information on the soil's composition, strength, compaction, permeability and other physical properties", enabling them to make informed decisions regarding the design of foundations and structures. Recommendations may include improving the soil (compaction, mixing with cement) or adapting the foundation to the existing soil (deep foundations in soft soils). Once the works have been completed, monitoring (inclinometers, piezometers) often continues to verify that the behaviour of the ground conforms to the predictions.

CHAPTER 3

Advanced applications of soil science

In modern geotechnical practice, it is essential to apply advanced techniques for analysing and mitigating ground risks, as well as innovative soil improvement methods, to ensure the stability and durability of structures. This chapter first addresses advanced geotechnical risks (liquefaction, settlement, expansive behaviour and collapsibility), describing their identification, effects and analysis criteria. It then presents the most modern soil improvement techniques, such as jet grouting, the use of geosynthetics, gravel columns and dynamic compaction, explaining their operating principles and applications. Finally, it illustrates some representative real-world cases where these phenomena and solutions have been observed or applied, reinforcing their practical relevance in civil engineering projects.

3.1 Geotechnical risk analysis

Geotechnical risk analyses require the identification of phenomena that may affect the integrity of structures. Among the most critical risks are seismic liquefaction, progressive subsidence, expansive soils and collapsible soils, each with its own specific causes and effects.

3.1.1 Seismic liquefaction

Liquefaction is a dynamic phenomenon that occurs in saturated non-cohesive soils (sand, gravel or non-plastic silt) during strong vibrations. Under seismic action, there is an increase in pore pressures and a consequent loss of the soil's effective strength. In the words of Geostru (2020), "liquefaction is a phenomenon that occurs in saturated non-cohesive soils... following a seismic swarm, reducing the soil's shear strength as pore pressure increases". This process can cause permanent deformations, local collapses or the formation of sand cones, with lateral displacements and foundation subsidence. For example, following the Maule earthquake (Chile, 2010), more than 180 sites with liquefaction were recorded up to 950 km from the epicentre.

To assess susceptibility to liquefaction, simplified methods (e.g. Seed & Idriss) are used, which compare the cyclic strength of the soil with the stresses induced by the earthquake. Key factors include the relative density of the soil, grain size distribution, fine-grained content, water table depth and the duration of the earthquake. Two main modes are distinguished: flow liquefaction (flow of liquefied soil on slopes) and cyclic softening (cyclic softening with limited deformation). In the Niigata earthquake (Japan, 1964), for example, liquefaction caused subsidence of up to 1.4 m in areas of alluvial deposits, with the partial collapse of buildings even at relatively low acceleration levels.

Mitigation measures include soil improvement techniques (compaction, vibro-flotation, gravel columns) to densify the ground, or drainage systems (vertical drains, drainage geosynthetics) to relieve pressure. Current regulations (Eurocode 8, Part 5, etc.) require susceptibility zoning and the design of foundations taking this risk into account.

3.1.2 Ground Subsidence

Subsidence is the progressive sinking of the earth's surface caused by various factors, particularly changes in the water table or in the composition of the subsoil. In simple terms, it is defined as “the progressive sinking of the ground due to the extraction of underground resources (water, oil) or other processes”. It often occurs when the overexploitation of aquifers leaves voids in compressible sediments, or due to the collapse of cavities (mining, karst). For example, the Basin of Mexico—rising above ancient lacustrine deposits—subsides by several tens of centimetres a year due to the pumping of groundwater.

The consequences include damage to foundations, cracks in structures and differential settlement. Assessment involves a combination of techniques such as extensive levelling surveys, GNSS/InSAR monitoring, consolidation tests and geomechanical modelling. In practice, empirical models and numerical analysis (finite element methods) are used to estimate curvatures and settlements. Critical subsidence has been documented in cities such as Jakarta, Mexico City and Phoenix, necessitating solutions involving abandonment or reinforcement. Consequently, when designing structures on susceptible ground, water extraction is usually restricted, surface drainage is improved and, in severe cases, deep foundations or soil stabilisation are employed

3.1.3 Expansive soils

Expansive soils contain clay minerals (mainly montmorillonite) that absorb water and swell, or shrink when they lose moisture. They are characterised by significant changes in volume in response to variations in water content. As can be seen in the expansion potential tables, montmorillonite can undergo very high increases in volume when submerged in water. Differential swelling beneath rigid footings generates uneven vertical thrusts, causing cracks, uplift and structural damage to buildings, pavements and pipes.

This risk is particularly prevalent in semi-arid climates and in poorly consolidated wind-blown or alluvial deposits. Identification is based on swelling index tests and standard geotechnical analysis (Atterberg limits, colloidal content). Mitigation measures include limiting infiltration (perimeter drains, impermeable barriers), designing foundations that transfer loads to deeper strata (piles or foundation slabs on non-expansive layers) and stabilising the soil (adding lime/cement to modify the structure). Studies show that damage caused by movement in light soils is largely attributable to swelling if the moisture cycle is not controlled.

3.1.4 Collapsible soils

Soils are referred to as collapsible if, when saturated, they undergo a sudden increase in volume and a sharp decrease in strength, causing their original structure to collapse. In their dry state, they are highly porous; when wetted, they lose the strength of their structure, which is based on surface tension and capillary forces, leading to sudden subsidence when the surface is loaded. They are common in poorly consolidated wind-blown or alluvial deposits in arid regions.

The resulting loss of volume can cause sudden failure in lightweight foundations. Diagnosing this requires specialised testing (double consolidation or oedometer tests under controlled humidity conditions). There is no single solution: intervention is on a ‘case-by-case’ basis. In general, measures include excavation and replacement with non-collapsible material, the installation of reinforcement elements (geosynthetics, micropiles) or the injection of stabilisers (cement or resins) to create more rigid structures that prevent sudden subsidence.

The literature indicates that once saturated, these soils can cause "severe damage" to infrastructure if precautions have not been taken.

3.2 Modern soil improvement techniques

To ensure the safety of construction works on problematic soils, advanced soil improvement techniques are employed. These include high-pressure grouting (jet grouting), the use of geosynthetics for reinforcement and drainage, stone columns for soil compaction and drainage, and deep dynamic compaction to densify loose soils. Their principles and typical applications are described below.

3.2.1 High-pressure grouting (jet grouting)

High-pressure grouting, or *jet grouting*, involves injecting jets of grout (cement or other binding agents) at high speed into the existing ground, breaking up and mixing the soil in situ. According to Víctor Yepes (2019), "high-pressure grouting... is a process that involves breaking up the soil..., mixing it and partially replacing it with a cementing agent (usually cement)", significantly improving the strength and reducing the permeability of the soil. Jet-grouting systems (single, double or triple fluid) allow for the creation of in-situ soil-cement columns, piles or walls with typical diameters of 0.5–1.0 m. This technique is applicable in soft soils (loose sand, silt, soft clay) where conventional methods would not work. Its applications include reinforcing weak foundations, waterproofing dams, stabilising slopes or improving the ground beneath tunnels. Jet grouting is also effective for filling cavities and preventing seepage, adapting to various soil conditions thanks to variations in the grout composition (cement, bentonite, etc.).

In addition to jet grouting, there are other low-volume injection techniques, such as the *injection of expanding resins* to lift deformed structures and the *injection of cement grouts* to improve granular soils, creating soil-cement nodules at depth. In general, grouting alters the original soil structure, creating new cohesive contacts or adding stiffness. Although costly, these methods allow intervention without excavation, which is suitable in confined spaces or on existing structures.

3.2.2 Geosynthetics

Geosynthetics are synthetic materials (e.g. geotextiles, geogrids, geomembranes, geocomposites) manufactured from polymers such as polypropylene or polyester, designed to address soil, drainage and stability issues. In civil engineering, they are used to reinforce, separate layers, filter solids, drain water and contain liquids. For example, non-woven geotextiles act as filters and geodrains, whilst woven geotextiles or geogrids are used as reinforcement in soft soils or on steep slopes.

These synthetic barriers and mesh offer significant advantages: they reduce the amount of conventional material required, facilitate rapid construction and enhance stability. According to SAI (2024), geosynthetics "reinforce soils, improve stability on slopes and distribute loads more effectively on roads". For example, for roads on soft clay, the installation of geotextiles separates the sub-base from the fill, prevents subsidence and improves drainage. In walls and slopes, geogrids allow for steeper and more stable gradients than with traditional methods (reducing the area required). Overall, geosynthetics provide modern, efficient and sustainable solutions for large-scale projects, particularly in challenging environments where other techniques would be unfeasible.

3.2.3 Gravel columns (stone columns)

Gravel columns are vertical inclusions of compacted granular material (gravel or a gravel-sand mixture) inserted into soft soil to reinforce it. Menard (2025) defines them as "flexible inclusions constructed by incorporating compacted granular material into the ground, forming a continuous column throughout the compressible layer". They are installed using vibroflotation: a vibrator excavates and compacts the gravel in successive passes, displacing the soil laterally without removing it. The typical diameter of a column is 0.6–0.8 m, although this may vary depending on the compressibility of the ground.

The effect combines mechanical reinforcement and drainage: the columns act as pathways for the evacuation of consolidation water and as piles that increase the shear strength of the surrounding soil. Among their advantages are accelerating the consolidation of soft soils (due to their draining nature) and providing greater rigidity to the structure as a whole. In particular, the technique is highly effective against seismic liquefaction: "by combining drainage and improved shear strength, gravel columns are an effective solution for anti-liquefaction treatment". They are used beneath shallow foundations (isolated or continuous footings) or in grids beneath distributed loads. The resulting improvement helps to reduce differential settlement and increase the soil's bearing capacity.

3.2.4 Dynamic compaction

Dynamic compaction is a technique for deep densification of granular soils and loose fill. It involves repeatedly dropping a heavy mass from a height (typically 10–30 metres) to generate pressure waves that compact the ground. Keller (2025) describes it as "a technique... that densifies and increases the bearing capacity of fill or granular and loose soils by utilising the free fall of a weight". The impact generates an initial compressive wave that raises the pore pressure and reduces the soil's temporary resistance, followed by a shear wave that fractures the granular structure and promotes ground settlement.

This technique is particularly useful for loose soils, sands and silts in recent fill. It improves the relative density, angle of friction and stiffness of the treated material, increasing load-bearing capacity and reducing subsequent settlement. Its common applications include the reinforcement of port yards, airports or heavy-load storage yards (for example, projects in Sonora, Mexico). Furthermore, dynamic compaction mitigates the risk of liquefaction in susceptible soils. As explained by Víctor Yepes (2016), this method "allows soils to be densified to great depths through the effect of high-energy impacts generated by the fall of a mass" and is applicable to both unsaturated soils and those below the water table. The design must take into account the distribution of impacts, energy and intervals between phases, based on preliminary tests and in-situ measurements (penetrometers, settlement monitoring) to ensure the quality of the treatment .

3.3 Real-world case studies

Some illustrative real-world cases are summarised below:

Maule Earthquake (Chile, 2010): The major 8.8-magnitude earthquake on 27 February caused soil liquefaction across a vast area (~950 km), affecting more than 180 sites from Valparaíso to Lake Llanquihue. Lateral displacement of embankments, subsidence in backfilled areas and numerous sand volcanoes were recorded. In soils at Nancagua and Retiro (Chile), laboratory tests revealed sands with a high content of fine particles (SC, ML) and a contractive behaviour, making them potentially liquefiable for accelerations >0.3 g. This study revealed that soils with a history of liquefaction may liquefy again in future earthquakes, highlighting the need for susceptibility maps and reinforcement measures in critical areas.

Niigata Earthquake (Japan, 1964): In June 1964, a Mw7.6 earthquake caused widespread liquefaction in the city of Niigata. Thousands of buildings tilted or sank on soft sandy deltaic deposits. There was local subsidence of up to 1.4 m, and the failure of foundations even under moderate accelerations demonstrated the danger of liquefaction. This historical case is frequently used as an academic reference for the devastating effects of liquefaction.

Subsidence in Mexico City: The Mexico City metropolitan area, built on clays and lacustrine sediments, suffers from chronic subsidence, which in some areas exceeds 50 cm per year. This is mainly due to intensive groundwater pumping, which compacts the soft clays. The subsidence has caused cracks in buildings and pipes, and has required solutions such as the injection of cement grout into deep boreholes (underpinning) and restrictions on pumping. This is an example of severe anthropogenic subsidence, typically found in megacities in arid climates.

Gravel column improvement (Central America): In recent projects in Latin America (for example, fuel tanks at the port of Corinto, Nicaragua), networks of gravel columns have been installed to reinforce soft soils and reduce settlement. As highlighted by Menard Latam, these features increase bearing capacity and accelerate the consolidation of silty soils with high water content. Case studies in Honduras and Panama demonstrate the effectiveness of this technique for building on land susceptible to liquefaction and subsidence.

CHAPTER 4

Environmental Geotechnics, Geophysical Methods and Site Monitoring

4.1 Environmental geotechnics in infrastructure projects.

Environmental geotechnics is a specialised branch of civil engineering that analyses the interaction between soil, groundwater and pollutants generated by human activities or industrial processes. This approach is essential in infrastructure projects, as it enables the assessment of the potential impacts of the ground on the works and vice versa, ensuring that construction solutions are compatible with the preservation of the environment. In this context, engineers must take into account factors such as soil contamination, fluid movement, the chemical behaviour of materials and the risk of hazardous substances leaching. By addressing these aspects, the likelihood of structural damage is reduced and compliance with current environmental regulations is ensured. Furthermore, environmental geotechnics facilitates informed decision-making throughout all phases of the project, from initial exploration to long-term maintenance.

4.1.1 Assessment of soil contamination

Soil contamination analysis is essential when developing sites previously occupied by industrial facilities, petrol stations, landfill sites or areas where accidental spills have occurred. This process involves taking deep soil samples to identify chemicals such as heavy metals, hydrocarbons, solvents and volatile organic compounds that may pose risks to human health and infrastructure. Characterising contaminated soil helps in selecting appropriate remediation methods, which may include techniques such as bioremediation, soil washing, physical encapsulation or in-situ chemical treatments. It is also necessary to consider the hydrogeological conditions of the area, as a contaminant can migrate into aquifers if not properly controlled. A comprehensive assessment requires integrating chemical analyses, laboratory tests and modelling of contaminant behaviour in different geotechnical scenarios.

4.1.2 Design of containment systems

The design of containment systems aims to prevent pollutants from spreading from facilities such as landfills, industrial plants or fuel storage facilities into the natural environment. To this end, impermeable geomembranes, bentonite layers, reactive barriers and drainage systems specifically designed to control water percolation are used. Each component plays a specific role, ranging from preventing vertical infiltration to reducing the lateral movement of leachate. In many cases, several protective layers are combined to form composite barriers that offer greater security against pressure changes or soil deformation.

It is also essential to carry out regular monitoring to ensure that the barriers continue to function properly as the hydrodynamic behaviour of the ground changes. Retaining structures have become a critical element of modern projects, particularly in densely populated urban areas.

4.1.3 Groundwater management

Groundwater management is a key component of environmental geotechnics, as construction works can significantly alter the natural balance of aquifers and internal flows within the subsoil. Deep excavations, basements, tunnels and underground foundations tend to alter pore pressures, which can lead to unexpected seepage or differential settlement if not properly controlled. To avoid risks, systems such as relief wells, waterproof screens and temporary pumping are implemented to stabilise hydraulic conditions during construction. The hydrogeological analysis must also consider the potential for internal erosion caused by upward flow, a phenomenon capable of compromising the stability of structures on sandy soils. Once the works are complete, the behaviour of the water table must be reassessed to ensure that no undesirable long-term impacts arise.

4.2 Geophysical methods applied to the characterisation of the subsurface

Geophysical methods enable information about the subsurface to be obtained without the need for extensive drilling or deep excavation, which is particularly valuable in projects covering large areas or requiring minimal disturbance to the ground. These methods utilise physical principles such as the propagation of seismic waves, electrical resistivity or the reflection of electromagnetic waves to identify changes in soil properties. Their use has increased significantly due to their efficiency and the ability to generate three-dimensional models of the subsurface that complement traditional geotechnical data. Furthermore, geophysical methods enable the detection of unstable zones, cavities, hard strata or water tables before drilling commences, thereby optimising resources and reducing costs. Combining them with targeted drilling and laboratory testing provides a more robust and reliable approach to the comprehensive characterisation of the ground.

4.2.1 Seismic refraction method

The seismic refraction method involves generating controlled seismic waves that travel through the ground and are recorded by sensors placed on the surface. As the waves pass through different materials, their speed changes depending on the stiffness and density of each layer, making it possible to identify deep layers without the need for excavation. This technique is particularly useful for detecting the position of bedrock, locating significant fractures and determining the thickness of sedimentary deposits. It is also used to assess areas susceptible to geological faults and to plan deep foundations for large-scale projects. Its main advantage is that it provides continuous profiles of the subsoil, unlike traditional boreholes which offer only point-specific information. However, the method requires suitable ground conditions and technical expertise to correctly interpret the recorded signals.

4.2.2 Electrical tomography

Electrical tomography is a technique widely used in geotechnical studies due to its ability to identify variations in the electrical resistivity of the subsoil through the controlled injection of current between electrodes. These variations enable the identification of saturated zones, clay deposits, natural cavities, aquifers or potential underground contamination. e data is processed using mathematical models that generate two-dimensional or three-dimensional images of the ground, facilitating the identification of internal structures without the need for excavation. Electrical tomography is useful in studies of unstable slopes, the design of special foundations and the exploration of areas where access is limited. Its accuracy depends on the configuration of the electrodes, the quality of the measurements and knowledge of the electrical behaviour of each type of soil. With these elements, it is possible to make more informed decisions for geotechnical design.

4.2.3 Ground-penetrating radar (GPR)

Ground-penetrating radar (GPR) is a geophysical method that emits electromagnetic waves into the ground and records the echoes reflected by changes in the soil's dielectric properties. This makes it possible to detect pipes, buried structures, voids, minor faults or interfaces between materials without the need for excavation. It offers high resolution compared to other geophysical methods, making it an ideal tool for urban surveys and for rapid diagnostics before opening trenches or carrying out excavations. It is also used on road surfaces to assess thickness, damaged layers or saturated areas that could affect the service life of the roads. Although its depth range may be limited by very clayey or saturated soils, GPR remains one of the most versatile and reliable tools for obtaining detailed information about the subsoil in a short time.

4.3 Geotechnical monitoring and deformation control

Geotechnical monitoring allows real-time observation of how the ground responds during construction and throughout the service life of a structure. Due to the variable nature of the soil, theoretical models often fail to accurately predict its behaviour; consequently, instrumentation becomes an indispensable complement to geotechnical design. By recording pressures, displacements and changes in the water table, anomalous behaviour can be detected before it poses a risk to the works. This approach is key in projects such as tunnels, slopes, deep excavations or large-scale foundations. Furthermore, the data obtained allows construction procedures to be adjusted or critical areas to be reinforced when measured values exceed established safety thresholds. Continuous monitoring improves design reliability and helps to enhance structural safety.

4.3.1 Instrumentation in excavations and structures

Geotechnical instrumentation includes equipment such as inclinometers, piezometers, load cells, extensometers and multi-parameter sensors, each designed to measure a specific variable relating to ground behaviour.

In deep excavations, inclinometers enable the identification of lateral ground movements that could compromise the stability of nearby walls or slopes. Piezometers, meanwhile, measure pore pressure in saturated soils, which is essential for preventing failures caused by liquefaction or hydraulic heave. Load cells enable the quantification of stresses in anchors or piles, whilst extensometers detect internal deformations that are not visible on the surface. The combined use of these instruments provides a comprehensive view of ground behaviour, improving the ability to anticipate problems and optimising mitigation strategies.

4.3.2 Satellite monitoring and remote sensing techniques

Satellite monitoring using technologies such as InSAR has revolutionised geotechnical monitoring by enabling the detection of millimetre-scale ground movements over large areas without the need to install equipment directly on site. This capability is particularly useful for studying subsidence, dam deformations, slope movements and cumulative urban deformations. In addition, the use of drones equipped with high-resolution cameras or LiDAR technology enables the generation of up-to-date three-dimensional models of hard-to-reach areas. These remote tools facilitate the continuous inspection of structures and exposed surfaces, reducing risks to personnel and providing accurate information for decision-making. Furthermore, the integration of historical imagery with recent monitoring data allows for the study of movement trends that could predict future failures.

4.3.3 Early warning systems

Early warning systems combine geotechnical instrumentation, automatic sensors and numerical models to issue warnings when measurements indicate abnormal behaviour. These systems are essential in high-risk areas such as unstable slopes, dams, operational tunnels and areas of intense seismic activity. By detecting sudden changes in displacement or pore pressure, the system can notify those responsible so that emergency measures can be implemented before a failure occurs. In urban contexts, these systems help to protect the population, preventing serious consequences. Furthermore, the information gathered allows predictive models to be refined and future geotechnical interventions to be improved. The combination of real-time monitoring and continuous analysis increases the resilience of structures in the face of unexpected events.

4.4 Integration of geotechnical engineering, geophysics and monitoring in modern projects

The integration of geotechnical engineering, geophysical methods and geotechnical monitoring provides a more comprehensive and accurate understanding of subsoil behaviour in modern engineering projects. By combining geophysical data with laboratory tests and real-time observations, sophisticated three-dimensional models are generated that can predict the ground's response under different loading conditions. This integration also enables the optimisation of foundation designs, reduces exploration costs and improves the early detection of potential risks. In large-scale projects, this methodology facilitates the planning of safer and more efficient construction phases, reducing the likelihood of unexpected failures. Furthermore, the coordinated use of these disciplines fosters a sustainable approach, minimising unnecessary interventions and promoting environmental protection.

CHAPTER 5

(Part 1): Advanced Geotechnical Design and its Application in Major Infrastructure Projects

5.1 Introduction to Advanced Geotechnical Design

Advanced geotechnical design is one of the most complex and critical areas within civil engineering, as it integrates an in-depth analysis of soil behaviour with structural, hydraulic, seismic and environmental criteria. Unlike conventional geotechnical studies, which focus primarily on characterising the ground and assessing its basic properties, advanced design considers more sophisticated interactions between infrastructure loads and the non-linear response of the subsoil under multiple scenarios. This approach allows for the evaluation of extreme conditions, such as sudden increases in pore pressure, long-term consolidation, time-dependent deformations, or physical-chemical changes in the soil induced by construction works. The purpose of this discipline is to optimise the safety and durability of complex structures such as dams, tunnels, deep foundations, industrial plants, cable-stayed bridges and mass transit networks. Its scope, therefore, goes beyond the limits of traditional geotechnics and is geared towards the comprehensive design of systems that depend directly on the behaviour of soil and rock.

Advanced geotechnical design requires a combination of multidisciplinary expertise and modern technological tools that enable the behaviour of the ground to be simulated using constitutive models. These simulations are based on laboratory and field data, and are adjusted to real-world conditions obtained through continuous monitoring. In this way, engineers can predict how the ground will behave under increasing loads, repetitive vibrations, water infiltration or prolonged consolidation processes. The analysis also includes soil-structure interaction, which involves assessing how stresses and strains change in both the ground and the structure over time. In short, this approach is essential for large-scale projects, where exaggerations or incorrect interpretations could result in failures with catastrophic consequences. The precision and depth required by this type of design justify the enormous technological and scientific development of modern geotechnics.

5.2 Fundamentals of advanced geotechnical analysis

The basis of advanced geotechnical analysis lies in the detailed study of the mechanisms governing soil behaviour under different loading conditions and states. Key concepts include the mechanics of porous media, consolidation theory, plasticity models, time-dependent behaviour, and the influence of pore pressure on soil stiffness. These concepts enable a precise interpretation of how a saturated or partially saturated soil responds to increases in stress, variations in moisture content, loading and unloading cycles, or changes induced by thermal processes. Understanding these phenomena enables the definition of key parameters used in numerical models, such as the modulus of elasticity, effective cohesion, angle of friction, consolidation parameters and stress-strain curves.

A key aspect of advanced design is the distinction between the drained and undrained behaviour of the ground. In saturated fine-grained soils, additional stresses are initially transferred to the pore water due to the material's low permeability, generating excess pressures that reduce shear strength and may lead to temporary failure. In contrast, in more permeable granular soils, the response is typically drained, allowing water to escape more rapidly and stresses to be transmitted directly to the soil skeleton. This distinction influences not only strength but also the mechanisms of deformation, settlement, consolidation and the overall stability of a structure. Correct interpretation of drained or undrained behaviour is essential for designing deep foundations, calculating differential settlements and analysing slope stability under seismic or static conditions.

Another key component of advanced analysis is the use of sophisticated constitutive models that capture the actual behaviour of soil under different stress levels. Unlike the simple linear models used in preliminary analyses, advanced models allow for the simulation of material hardening and softening, anisotropic behaviour, dependence on confining stress, and changes in stiffness associated with repetitive loading cycles. Models such as Mohr-Coulomb, Hardening Soil, Modified Cam-Clay and non-linear variants are used in specialised software such as PLAXIS, FLAC, GeoStudio and ABAQUS to reproduce complex scenarios. These models require parameters carefully calibrated from advanced triaxial tests, edometric tests, in-situ measurements and laboratory analyses using high-precision equipment. The quality of the model depends directly on the quality of the data, making a rigorous and well-planned geotechnical investigation programme essential.

5.3 Constitutive models for soils and rocks

Constitutive models represent the relationship between stresses and strains in soils and rocks. In advanced geotechnical design, these models enable the simulation of the actual behaviour of materials under conditions that cannot be simplified using linear assumptions. There are different categories of constitutive models, ranging from the simplest to those that include multiple variables and complex parameters. Linear elastic models are the most basic and assume that the relationship between stress and strain is proportional within the material's elastic range. Although they are useful for preliminary analyses or for materials with essentially rigid behaviour, their use is limited in major projects where it is necessary to predict strains and failures with precision.

Elastoplastic models, on the other hand, take into account the material's ability to undergo permanent deformation once the elastic limits are reached. Within this category, models based on the Mohr-Coulomb criterion stand out; these define shear failure in terms of cohesion and the effective angle of friction. This model is widely used due to its simplicity and because it represents the behaviour of granular and cohesive soils reasonably well. However, its application is limited when it is necessary to consider variations in stiffness with the stress level. To overcome this limitation, more advanced models such as the Hardening Soil model are employed, which accounts for the hardening of the material due to changes in the stress state, allowing for the simulation of non-line e deformations and the effect of confinement.

Another important model in advanced geotechnical analysis is the Modified Cam-Clay model, used primarily for normally consolidated clays. This model incorporates concepts from porous media mechanics and allows the simulation of both consolidation and the dilatancy and contractility of the material under different load levels. Visco-plastic models are also used to analyse the time-dependent behaviour of soft soils or materials that exhibit creep. In the case of rock, constitutive models must incorporate the brittle behaviour of the material, as well as the presence of joints, fractures and planes of weakness. Models such as Hoek-Brown allow the simulation of failure in rock masses using parameters that represent the intrinsic strength of the material and the quality of the rock mass. Taken together, advanced constitutive models provide engineers with tools capable of predicting ground behaviour under real-world and extreme conditions.

CHAPTER 6

Special Foundation Engineering and Soil-Structure Interaction Systems

6.1 Introduction to Special Foundations

Special foundations represent an advanced category within civil engineering, focused on the design and construction of systems capable of transmitting loads to deep strata or interacting effectively with problematic soils. Unlike conventional shallow foundations, which rely primarily on the bearing capacity of the surface soil layers, these systems utilise more complex transfer mechanisms that enable them to support heavy buildings, critical infrastructure and structures subjected to significant dynamic loads. Their applications include projects such as skyscrapers, cable-stayed bridges, dams, industrial buildings, power towers, ports and offshore platforms. Modern engineering has driven the development of specialised techniques that combine deep drilling, grouting, precast elements and seismic behaviour control devices to ensure structural stability in challenging geotechnical conditions.

These foundations must not only withstand considerable vertical loads, but also resist horizontal forces, torsion, vibrations, thermal expansion and fluctuations in the water table. The geological conditions of the site directly influence the choice of system, as soft soils, collapsible strata, highly compressible clays or heterogeneous materials require solutions that allow forces to be distributed through several simultaneous mechanisms. In practice, the design of special foundations involves a comprehensive analysis of the interaction between the subsoil and the structure throughout its entire service life. This interaction is dynamic and depends on hydraulic changes, consolidation, deformations due to static loads, increases in cyclic stresses and environmental changes. Therefore, each project requires a bespoke approach that takes into account the specific characteristics of the ground and the structural demands of the project.

6.2 Fundamentals of load transfer in deep foundations

Deep foundations are based on the principle of transferring loads from the structure to stiffer or more capable layers of the subsoil via vertical or inclined elements. Load transfer occurs via two fundamental mechanisms: lateral friction and tip resistance. Both mechanisms operate simultaneously, but their relative contribution depends on the soil type, the material of the pile or micropile, the construction method and the embedment depth. In soft soils with high compressibility, lateral friction may dominate because tip resistance is low. Conversely, in dense or rocky strata, tip resistance tends to be the main component of the system. This balance between mechanisms is essential for predicting the behaviour of the pile under vertical and lateral loads, as well as under dynamic conditions.

The study of these mechanisms requires an analysis of the relative stiffness between the pile and the surrounding soil. When the pile is considerably stiffer than the soil, the load distribution tends to be concentrated near the tip, whereas a more flexible pile distributes stresses gradually along its shaft. The interaction of these properties is assessed using reaction coefficient models, load-settlement curves and three-dimensional finite element analysis. Furthermore, the load-bearing capacity of the system depends on factors such as group efficiency when using pile clusters, soil consolidation following installation and the influence of lateral displacement induced during driving or drilling. Each of these elements is essential to ensure that the system is stable and safe.

6.3 Displacement piles and bored piles

Displacement piles and bored piles represent two main categories within deep foundations, each with particular characteristics that make them suitable for different ground conditions. Displacement piles, such as driven or pre-excavated piles with lateral expansion, compact the surrounding soil during installation. This process can improve the geotechnical properties of the ground by reducing porosity and increasing the available lateral friction. Steel, prestressed concrete or timber piles fall into this category and are used in projects requiring rapid installation and high load-bearing capacity. Their main advantage lies in the improvement of the surrounding ground, although their use can generate significant vibrations that limit their application in sensitive urban areas.

On the other hand, bored piles are constructed by excavating a cylindrical shaft which is subsequently filled with concrete and, in many cases, reinforced with steel. This type of pile allows work to be carried out in soils containing isolated rocks, hard strata, pockets of rock, or conditions where driving piles would be impractical. Furthermore, they allow for greater diameters and depths and are commonly used in bridge foundations, multi-storey columns or structures with high horizontal loads. Their installation requires strict control of the stratigraphic profile, the use of bentonite slurries or stabilising polymers, and the appropriate design of the concrete mix to prevent segregation. Compared to displacement piles, they generate less vibration, but do not improve the surrounding ground, so their bearing capacity depends exclusively on direct contact between the concrete and the soil.

6.4 Inclined piles and their contribution to overall stability

Unlike vertical piles, which are mainly used to resist gravitational loads, inclined piles are driven into the ground at a specific angle to control horizontal stresses, torsional forces and lateral thrusts. These systems are essential in structures subjected to significant horizontal forces, such as bridges, quays, buildings exposed to strong winds and platforms subject to unexpected changes in water level. The inclination of the piles allows the lateral friction and tip resistance components to act against these forces, creating a stable system capable of redistributing stresses to deeper layers. This mechanism enables the structure to withstand critical events, such as earthquakes or the impact of sea currents, without experiencing excessive displacement.

The design of inclined piles requires three-dimensional modelling to assess their behaviour at different angles and depths. Soil-pile interaction varies depending on the degree of inclination; therefore, lateral and e deformations, induced axial stresses, and potential stress concentrations at specific points must be analysed. Their installation also presents particular challenges, as it requires specialised equipment to ensure the element is correctly oriented without compromising its structural integrity. Furthermore, factors such as the potential reduction in capacity due to inclined excavation, alignment control and the influence of the pile group on load distribution must be considered. Nevertheless, when correctly designed and installed, inclined piles provide exceptional stability even under complex geotechnical conditions.

CHAPTER 7

(Part 1): Soil Dynamics and Vibration Control in Civil Engineering

7.1 Introduction to soil dynamics

Soil dynamics is a fundamental branch of modern geotechnical engineering dedicated to the study of soil behaviour when subjected to dynamic loads, vibrations or time-varying stresses. Unlike traditional static analyses, soil dynamics takes into account the influence of frequency, wave propagation velocity, the dynamic stiffness of the soil, and the interaction between the structure and the soil under transient conditions. These conditions arise in numerous infrastructure projects that must cope with disturbances such as earthquakes, controlled explosions, the impact of heavy machinery, vibrations generated by road or rail traffic, and cyclic loads from industrial equipment. The study of these phenomena is particularly relevant in densely populated urban areas, where vibrations can affect buildings, utility networks and the health of occupants.

Dynamic analysis enables us to understand how the ground absorbs, reflects, transmits or amplifies the waves that propagate through its various layers. Each type of soil responds in a particular way to dynamic effects depending on its density, stiffness, saturation, internal structure, degree of cementation and level of confinement. For example, loose sands tend to lose stiffness under repetitive loads, whilst highly plastic clays may exhibit complex viscoelastic responses that depend on time. The soil's capacity to withstand cyclic stresses also determines its susceptibility to hazardous phenomena such as liquefaction, dynamic settlement or progressive softening. Soil dynamics has therefore become an indispensable discipline for ensuring the safety and functionality of structures in environments exposed to variable stresses.

7.2 Fundamentals of seismic waves in soils

Seismic waves are one of the most significant phenomena in the dynamic study of the ground, as they enable the analysis of how stresses are transmitted through the layers of the subsoil. These waves are classified into two main categories: internal waves and surface waves. Internal waves include P-waves (primary) and S-waves (secondary), which travel through the ground at different speeds and in different directions. P-waves propagate through compression and expansion; they are the fastest and the first to arrive during a seismic event. S-waves, on the other hand, travel through shear deformation and cannot propagate in fluids, so their speed depends directly on the soil's shear stiffness. The difference in the speeds of these two waves allows dynamic properties such as the shear modulus and soil density to be determined.

Surface waves, notably Rayleigh and Love waves, propagate mainly near the Earth's surface and are responsible for most structural damage during an earthquake. Rayleigh waves produce elliptical movements in the vertical plane, whilst Love waves generate horizontal displacements of a transverse nature. The intensity of these waves is strongly influenced by the thickness, composition and stiffness of the surface layers . In soft soils, surface waves can undergo significant amplification due to the low stiffness of the material, resulting in more intense movements compared to denser or rocky soils. A detailed analysis of wave propagation enables the design of resilient structures and the assessment of the ground's seismic behaviour under different scenarios.

7.3 Dynamic properties of soils

The dynamic properties of the ground describe its behaviour under cyclic loads or vibrations, and form the basis for any seismic or dynamic response analysis. Among the most important properties are the dynamic shear modulus, the damping ratio, frequency-dependent stiffness, and material degradation curves. The dynamic shear modulus represents the stiffness of the soil under small deformations and is calculated using specialised tests such as column resonance, Bender elements shear wave tests or cyclic triaxial tests. This parameter is usually much higher than the shear modulus obtained under static conditions, especially when deformations are small. The dependence of the dynamic modulus on the amplitude of deformation allows us to identify how the soil behaves during the initial, intermediate and severe phases of a dynamic event.

Damping, on the other hand, is the material's ability to dissipate energy during a loading cycle. Soils with a high clay content or a cemented structure typically exhibit higher levels of damping, whilst dense granular soils show a lower capacity to dissipate energy. Dynamic damping directly influences the amplitude of vibrations and the response of structures under periodic loads. Degradation curves, which relate the reduction in shear modulus to the increase in deformation, allow the non-linear behaviour of the material under severe loading to be characterised. These properties are essential for designing foundations, evaluating soil-structure interaction and estimating settlements induced by vibrations or earthquakes.

7.4 Dynamic laboratory and field tests

Dynamic tests provide fundamental parameters for the advanced analysis of the seismic or vibratory behaviour of the ground. Among the most commonly used laboratory tests is the cyclic triaxial test, in which a repetitive load is applied to simulate seismic conditions, enabling the determination of parameters such as dynamic stiffness, damping and susceptibility to liquefaction. Another essential test is the resonant column test, which applies small dynamic loads to cylindrical samples to measure their response at specific frequencies. This test is useful for characterising soil behaviour under low deformations, a condition commonly encountered in vibrations generated by machinery or traffic. In granular soils, the shear wave test using bender elements allows the propagation velocity of S-waves to be measured directly in saturated or partially saturated samples.

In the field, geophysical methods such as MASW (Multichannel Analysis of Surface Waves), ReMi (Refraction Microtremor) and down-hole techniques enable the determination of wave velocities that help define dynamic subsurface profiles. These methods are particularly useful in areas where obtaining undisturbed samples is difficult or where ground conditions are highly heterogeneous . The integration of laboratory and field tests enables the construction of accurate dynamic models that adequately represent soil properties under seismic or vibratory conditions. In critical projects such as bridges, dams, tunnels and high-rise buildings, these tests are essential to ensure that the design responds adequately to dynamic demands and reduces the risk of structural failure.

CHAPTER 8

(Part 1): Tunnels, Underground Excavations and Ground Behaviour in Deep Excavations

8.1 Introduction to Tunnel and Deep Excavation Engineering

Tunnel engineering is one of the most sophisticated branches of civil engineering, due to the inherent complexity of the behaviour of the subsoil when its natural equilibrium is disturbed at considerable depths. Unlike other surface works, where loads can be widely distributed and the ground can be reinforced with relative ease, tunnels and deep excavations generate three-dimensional changes in the stress state that require meticulous control from planning through to operation. In this type of infrastructure, stability depends not only on the strength of the soil or rock, but also on factors such as the initial confinement of the rock mass, the progression of the construction process, groundwater pressure and the loss of volume occurring around the excavation. Consequently, underground works require advanced analyses that integrate geotechnics, geology, hydrogeology, geophysics and high-precision three-dimensional numerical models.

The importance of this discipline is growing alongside the rise of densely populated cities and the need for underground infrastructure capable of carrying water, energy, mass transit and communication systems without affecting the surface. The construction of tunnels beneath buildings, roads, rivers or fractured rock formations involves significant risks that must be addressed using specialised excavation methods and support systems designed to adapt continuously to ground conditions. Consequently, the geotechnical approach cannot be limited to static values or traditional parameters, but requires an understanding of the subsoil's deformation behaviour, the redistribution of stresses, and the influence of tunnel advance on the surrounding environment. This need has driven the development of technologies such as balanced-pressure tunnel boring machines, sequential excavation methods and ground freezing techniques, which enable work to be carried out under conditions that were considered impossible just decades ago.

8.2 Fundamentals of soil behaviour around deep excavations

When a deep excavation is carried out, the surrounding ground undergoes a process of distortion that completely alters the previous balance between natural stresses and the internal strength of the materials. The partial or total removal of soil at a given point causes a redistribution of vertical and horizontal stresses, forcing the ground to deform towards the excavated area. This phenomenon, known as convergence or movement towards the void, occurs in both soils and rock, although its magnitude and form depend on the stiffness of the material, the presence of fractures, the water table, the rate of advance and the support system used. In soft soils with low stiffness, deformations are usually more pronounced and require immediate shoring measures, whereas in competent rock, its self-supporting capacity can be utilised before installing permanent reinforcements.

The behaviour of the ground is not uniform around the excavation, but rather follows patterns defined by the shape and size of the tunnel. Typically, the highest concentration of stresses occurs at the crown, where the material tends to fracture or relax depending on the quality of the rock mass. The side walls experience alternating tensile and compressive stresses, whilst the base is subjected to combined effects of thrust and potential hydraulic uplift if there is rising water pressure. These interactions make it necessary to use models that represent the three-dimensional stress state to predict anticipated movements, surface settlements and the soil's response to progressive changes in the support. The analysis must also take time into account, due to phenomena such as consolidation in saturated soils, creep in sensitive clays or relaxation in rocks with viscoelastic behaviour.

8.3 Tunnel excavation methods and their relationship to geotechnical engineering local

The choice of excavation method is one of the key factors determining ground behaviour during tunnel construction. The decision is based on the type of soil or rock, the depth of the tunnel, the presence of groundwater, the distance to sensitive buildings or structures, and the geometry of the project. In competent rock masses, the Drill & Blast method is commonly used, which involves drilling and controlled blasting to fragment the material and remove it with excavators. This method requires strict control of the drilling pattern and blasting energy, as a poorly designed blast can induce additional fractures that compromise the stability of the tunnel. In good-quality rock, however, it allows for rapid progress and efficient performance.

In soft ground or urban environments, mechanised methods such as tunnel boring machines (TBMs) are preferred, as they can operate using different ground pressure balance systems. Equilibrium pressure balance (EPB) TBMs control the pressure in the excavation chamber by using a mixture of excavated material and additives that act as a temporary support against the surrounding ground. On the other hand, slurry TBMs use bentonite slurry to stabilise the face and control seepage. Both systems allow work to be carried out beneath rivers, in saturated areas or in highly permeable soils without causing subsidence or sudden ground flow towards the machine. In shallow tunnels or those with variable cross-sections, the NATM (New Austrian Tunnelling Method) is applied, which uses sequential excavation, flexible support and continuous monitoring to adapt the support according to the actual response of the ground. This procedure is based on the principle that the surrounding rock mass or soil forms part of the structural system and must be utilised as such.

8.4 Interaction between excavation, the water table and hydraulic pressures

The water table plays a crucial role in the stability of a deep excavation, as the presence of water alters the effective strength of the ground and generates pore pressures that can lead to catastrophic failure if not properly controlled. When the excavation is below the water table, hydraulic pressure tends to push material into the tunnel, causing seepage, internal erosion and potential siphoning. In sandy soils, these effects can trigger sudden face failures or even widespread collapses. For this reason, underground works require control systems such as pumping wells, grouting and watertight injec , ground freezing or retaining screens to limit the migration of water into the excavation.

Groundwater level management must take into account not only the immediate stability of the tunnel, but also the impact that a prolonged drop in water levels may have on the surface. A fall in the groundwater level can lead to consolidation and differential settlement, affecting nearby buildings, pipes and road surfaces. In clay soils, this drop can result in significant contraction that compromises existing infrastructure, even where there is no direct contact with the tunnel. Consequently, drainage and pumping systems must be carefully designed, monitored in real time and adjusted as excavation progresses. In some projects, the preference is to avoid any disturbance to the groundwater, and balanced-pressure tunnel boring machines are chosen to maintain the hydraulic balance, allowing safe excavation beneath high water tables without affecting the surrounding environment.

CHAPTER 8 (Part 2): Tunnels, Underground Excavations and Ground Behaviour in Deep Excavations

8.5 Primary support systems and their interaction with the rock mass

Primary support constitutes the first line of defence against movements induced by excavation and is essential for ensuring the immediate stability of the tunnel. A key feature of these systems is that they must adapt to the initial deformations of the ground, allowing for controlled interaction between the rock mass and the lining. Among the most commonly used elements are shotcrete, anchor bolts, steel trusses, reinforcement mesh and temporary struts. Each fulfils a specific function depending on the geomechanical properties of the ground and the tunnel geometry. For example, in fractured rock masses, the use of bolts allows potentially unstable blocks to be 'stitched' together, whilst shotcrete provides an initial layer of rigidity that prevents rockfalls. The choice of support depends on the advance rate, the construction method and the degree of support required by the ground.

The interaction between the primary support and the ground is based on the concept of confined convergence, whereby the rock mass tends to deform towards the excavation and the support system limits these deformations to within acceptable limits. This combined behaviour allows a natural equilibrium to develop between the load-bearing capacity of the rock mass and the stiffness of the support. In soft soils, for example, it is necessary to install the support immediately after each advance to prevent deformations from increasing and compromising the stability of the tunnel. In contrast, in competent rock with good cohesion, some deformation may be allowed before applying the support, taking advantage of the rock mass's self-supporting capacity. The design of the primary support must also consider the presence of water, thermal conditions, the setting time of shotcrete, and the durability of the system against potential dynamic loads during the tunnel's service life.

8.6 Secondary support systems and permanent linings

The secondary lining represents the final stage of the tunnel's structural support and is designed to withstand permanent loads, hydrostatic pressure variations, long-term deformations and adverse environmental conditions. A permanent lining may consist of in-situ cast reinforced concrete, precast segments, high-strength shotcrete or composite structures with intermediate waterproof membranes. Its function is to stabilise the tunnel throughout its life cycle, providing structural strength, watertightness and durability. In tunnels constructed using a tunnel boring machine, the final lining is generally formed by rings of precast segments that are assembled immediately after the machine has advanced. These segments are designed to withstand external ground pressures, vertical surface loads and possible rises in the water table during extreme events.

In projects involving conventional excavation, the permanent lining is designed taking into account the convergence achieved during the first phase, which allows the thickness of the lining to be optimised and ensures efficient use of materials. In soils susceptible to chemical attack, additives and protective membranes are incorporated to prevent concrete degradation and avoid seepage over time . Furthermore, in seismic zones, linings must be capable of withstanding cyclic loads and differential movements without losing their integrity. In some cases, expansion joints, special reinforcements or flexible sections are incorporated to dissipate energy during seismic events. The durability of the lining also depends on the quality of construction, concrete curing, crack control and regular maintenance, particularly when the tunnel forms part of mass transit networks or critical infrastructure.

8.7 Waterproofing methods and leak control

One of the most significant challenges in underground projects is controlling the flow of water into the excavation. Hydrostatic pressure can cause seepage that affects the safety and durability of the tunnel, damages the linings and reduces the effectiveness of the support systems. To prevent this seepage, waterproof membranes, cement grouting, chemical injections, frozen ground barriers and bentonite barriers are used to seal the perimeter of the excavation. Waterproof membranes are particularly useful in tunnels constructed using TBMs, as they can be installed between the segments and the ground, preventing water from entering the interior. In tunnels excavated using conventional methods, grouting carried out before or after tunnelling allows cracks, fractures or permeable zones identified during geotechnical characterisation to be sealed.

The design of a waterproofing system must take into account not only water pressure but also the chemical aggressiveness of the environment and the expected service life of the tunnel. In environments where acidic water or dissolved minerals are present, seepage can accelerate the corrosion of reinforcement or cause concrete deterioration. For this reason, chemically resistant membranes are used and internal drainage systems are designed to capture and remove any residual seepage. Water flow control requires continuous monitoring using piezometers, inspection drains and pressure control systems. In exceptional events, such as floods or failures in nearby underground networks, the tunnel must have redundant drainage systems capable of handling higher-than-normal flow rates.

Errors at this stage can lead to delayed failures that compromise the operation of the infrastructure, even years after its construction.

8.8 Soil-structure interaction in urban tunnels

Tunnels built in urban areas present additional challenges due to the proximity of buildings, pipes, electrical networks, road surfaces and public transport. Excavation alters the stress state of the ground, generating settlements that can be transmitted to nearby structures. Although these settlements are usually small, even minimal values can affect shallow foundations, old buildings, retaining walls, or sensitive underground utilities. To prevent such damage, it is essential to accurately calculate expected displacements, use three-dimensional numerical models, and design mitigation measures such as compensatory grouting, ground improvement, jet grouting, artificial freezing, or temporary reinforcement with protective piles or micropiles. The key to success in urban construction lies in accurate prediction and continuous monitoring as the tunnel advances.

Soil-structure interaction depends on the type of building concerned, its age, structural rigidity and type of foundation. Some structures can tolerate uniform settlement without significant consequences, whilst others are extremely sensitive to movements of just a few millimetres. In this context, engineers must assess the susceptibility of each building and establish alert levels that allow for intervention before displacements exceed permissible limits. Modern monitoring techniques, such as inclinometers, extensometers, laser sensors and systems based on satellite interferometry, enable the detection of minute movements and immediate action. Furthermore, coordination with structural engineers, architects and local authorities is vital to ensure that tunnel advance does not compromise urban integrity or user safety.

CHAPTER 8 (Part 3): Tunnels, Underground Excavations and Ground Behaviour in Deep Excavations

8.9 Techniques of ground improvement applied to underground works

Ground improvement is an essential component of tunnel engineering, particularly when working in soils with low strength, high permeability or a high risk of deformation. These techniques aim to modify the properties of the soil to increase its stiffness, reduce its permeability, improve its stability or control deformation during tunnel excavation. Among the most commonly used methods are cement grouting, chemical grouting, in-situ compaction, jet grouting, thermal stabilisation and artificial ground freezing. Each of these techniques is applied depending on ground conditions, the depth of the excavation, the presence of water and the type of structure to be built. In saturated soft soils, for example, jet grouting allows the creation of soil-cement columns that act as a barrier and support, significantly reducing the likelihood of excessive settlement or seepage.

On the other hand, chemical grouting is useful when it is necessary to minimise soil permeability without significantly altering its natural structure. This method uses compounds such as silicates, acrylic resins or polyurethanes, which react chemically to form gels or solids within the soil pores. Artificial ground freezing is a more specialised technique used in tunnels beneath high water tables or in extremely unstable soils. It involves installing pipes through which a refrigerant fluid circulates, creating a frozen block around the excavation that acts as a temporary support and an impermeable barrier. Although this technique is costly, it allows for the management of geotechnical conditions where other methods would be insufficient or unsafe. The success of ground improvement depends on proper design, controlled execution and continuous monitoring to verify that the resulting properties meet the established parameters.

8.10 Methods for controlling surface settlement

During tunnel construction, particularly in urban areas, one of the most critical issues is the occurrence of ground settlement. This settlement results from volume loss around the excavation, the convergence of the ground towards the tunnel, and the subsequent redistribution of stresses. In cohesive soils, settlement can develop slowly due to consolidation processes, whilst in granular soils it usually occurs immediately. To control these effects, techniques such as counter-injection, ground pre-support, the use of balanced-pressure tunnel boring machines, and the reduction of the void generated during tunnelling are employed. Compensatory grouting, for example, allows controlled volumes of expansive material to be introduced above the tunnel to counteract downward deformation. This method requires millimetre precision, as any excess could cause unwanted uplift and compromise the stability of surface structures.

Another common strategy is ground pre-support, which involves reinforcing the rock mass prior to tunnelling using methods such as micropile umbrellas, pre- r grouting, or the installation of rigid frames. This method is used particularly in areas where the tunnel is at shallow depth and settlement could affect historic buildings, transport networks or critical infrastructure. In mechanised excavations, the pressure exerted by the tunnel boring machine at the excavation face and in the shield chamber helps to control deformations, provided that an appropriate balance is maintained between the support pressure and the ground resistance. Incorrect management of this pressure can cause sudden subsidence or excessive deformations that compromise the safety of the project. For this reason, settlement control is one of the most closely monitored aspects in urban tunnel construction.

8.11 Geotechnical monitoring in underground works

Geotechnical monitoring plays a vital role in the early detection of movements, deformations or changes in ground stability during the construction and operation of a tunnel. This monitoring is carried out using instruments such as inclinometers, extensometers, piezometers, load cells, convergence gauges and pressure sensors at the tunnel face. The installation of these instruments provides real-time information on the behaviour of the rock mass and the progress of the works, facilitating decision-making at critical moments. For example, a sudden increase in pore pressure may indicate a risk of siphoning or loss of stability, whilst excessive deformations may signal that the support system is not functioning properly. Monitoring allows for the correction of operating conditions, the adjustment of excavation parameters or the reinforcement of the support system before the situation escalates into a major problem.

The interpretation of monitoring data requires a thorough understanding of geotechnical behaviour, as measured values can have multiple causes and must be analysed within their context. Furthermore, monitoring is essential in projects where public safety depends on the proper functioning of the tunnel, such as underground railways, underground roads or high-capacity water mains. In recent years, the use of automated systems and remote technologies has revolutionised geotechnical monitoring. Methods such as satellite interferometry (InSAR), three-dimensional laser systems and fibre-optic sensors enable continuous, high-precision measurements to be obtained without interrupting the construction process. The integration of these systems with predictive models facilitates the early detection of anomalies and the immediate implementation of corrective measures.

8.12 Tunnels in fractured rock and weak strata

Tunnels built in fractured rock present challenges that require highly specific support and excavation strategies. The presence of discontinuities, faults, joints, planes of weakness and zones of alteration significantly reduces the strength of the rock mass and creates risks of rockfalls, blockages and unexpected movements. In such cases, geological and geomechanical characterisation is essential to determine the orientation, opening, filling and roughness of the fractures. This information enables the modelling of tunnel stability and the design of reinforcement systems to prevent rockfall or slippage along critical planes. Rock bolts and anchors are essential tools for 'stitching' the rock mass together and joining blocks that might otherwise move independently. Furthermore, shotcrete and welded wire mesh help to control surface rockfalls and stabilise altered zones.

In extremely weak strata, such as weathered shales, overconsolidated clays or highly altered rocks, it may be necessary to use special techniques such as short-step excavation, the use of micropile umbrellas or the installation of rigid support immediately after each advance. These materials tend to deform rapidly due to their low stiffness and cohesion, which requires strict control of the excavation rate. In some cases, even small vibrations can trigger significant deformation, so the use of heavy machinery or explosives must be restricted. The behaviour of weak rock masses may include creep, plasticisation and time-dependent movements, phenomena that must be taken into account in the design of the final lining and in the assessment of the tunnel's service life.

CHAPTER 8 (Part 4): Tunnels, Underground Excavations and Ground Behaviour in Deep Excavations

8.13 Excavations in expansive, collapsible and highly compressible soils

Underground excavations carried out in expansive soils present one of the greatest challenges in tunnel engineering, due to the significant changes in volume that these materials undergo when their moisture content varies. This type of soil, commonly associated with clay minerals such as montmorillonite, can increase in volume considerably when absorbing water, generating internal pressures capable of deforming linings, displacing primary supports and compromising the stability of the tunnel. During excavation, a reduction in confinement can accelerate moisture absorption, causing differential swelling that affects the cross-sectional shape and increases stresses at specific points in the lining. To mitigate these phenomena, reinforced waterproofing systems, perimeter drains and high-rigidity linings are used to reduce deformation induced by swelling.

On the other hand, collapsible soils exhibit the opposite behaviour, as they are characterised by a metastable internal structure that undergoes sudden reductions in volume when subjected to loads or when it comes into contact with water. In tunnels constructed through these materials, even minimal seepage can trigger a local collapse that spreads rapidly towards the excavation face, causing subsidence, longitudinal deformations and loss of support. To prevent such events, techniques such as stabilising pre-injections, sequential excavation with immediate support and strict control of the water table are employed. Highly compressible soils require a different approach, as their deformations are time-dependent and may continue to develop decades after the works have been completed. Consequently, linings must be designed to withstand differential settlement without losing watertightness or structural strength.

8.14 Tunnels in seismic zones and dynamic soil analysis

The construction of tunnels in seismic zones requires a detailed analysis of the dynamic behaviour of the ground and of soil-structure interaction during a seismic event. Unlike surface structures, which amplify or attenuate ground motion depending on their stiffness and shape, tunnels tend to behave as buried structures that follow the deformations of the

. This means that their safety depends on the rock mass's ability to deform without generating excessive stresses in the lining. Seismic analysis must consider the propagation of S- and P-waves, the dynamic stiffness of the material, potential liquefaction of the surrounding environment, and the difference in response between undisturbed ground and ground affected by excavation. In saturated soils with a high susceptibility to liquefaction, an earthquake can cause abrupt losses of stiffness that completely compromise the temporary support of the tunnel.

Seismic design of tunnels utilises advanced numerical models that simulate seismic waves of varying intensities and directions. These models assess axial deformations, cross-sectional ovalisation, torsion and accumulated stresses at joints and connection elements. In tunnels lined with precast segments, the joints must be designed to allow relative displacement without compromising watertightness or structural integrity. In monolithic linings, the reinforcement layout must ensure that the concrete can withstand repetitive loading cycles without developing critical cracks. Likewise, access points, ventilation shafts and auxiliary galleries must be designed to avoid stress concentrations that could cause local failures. Seismic engineering applied to tunnels is based on a combination of experimental results, historical seismic records and detailed numerical analyses that reproduce realistic deformation conditions.

8.15 Excavation beneath existing buildings and mitigation measures

When a tunnel must be constructed beneath existing buildings, the safety of those structures becomes the primary focus of the construction process. Excavation alters the stress state of the ground and causes settlement that can be transmitted to shallow or deep foundations. To minimise these effects, specialised techniques are employed, such as protective jet grouting, compensatory micropiles, low-pressure grouting and temporary structural reinforcement in vulnerable buildings. The aim is to redistribute the building's loads evenly and prevent settlement from concentrating at specific points. Initial calculations must include a detailed analysis of the type of existing foundation, the structure's stiffness, its age and its history of deformation to predict its response to the movements generated by the tunnel.

Continuous monitoring of nearby buildings is essential to ensure their structural integrity. Crack gauges, survey prisms, laser sensors and extensometers are installed to record movements with millimetre precision. These instruments enable changes in behaviour to be detected and emergency measures to be triggered when deformations exceed established limits. Preventive measures also include modifying the tunnel boring machine's operating parameters, adjusting the pressure in the excavation chamber, reducing the advance rate, or carrying out compensatory grouting in real time. In extreme cases, it may be necessary to temporarily shore up the structure or reinforce it using steel frames or auxiliary foundations. The key to success in these projects lies in multidisciplinary coordination between geotechnical engineers, structural engineers, architects and contractors, as well as in the implementation of reliable predictive models and robust monitoring systems.

8.16 Excavation methods for large-diameter tunnels

Large-diameter tunnels, such as those used for mass transit systems, deep drainage works and high-capacity water mains, require highly specialised construction methods. EPB and slurry tunnel boring machines are widely used in these projects due to their ability to control internal pressures and minimise ground settlement. However, their operation becomes more complex as the tunnel diameter increases, as face stability, stress distribution and the quality of the lining depend on precise control of parameters such as support pressure, advance rate and the consistency of the excavated material. Large- -diameter machines typically incorporate integrated monitoring systems that allow real-time adjustment of the pressure in the excavation chamber, the rotational torque of the cutter head, the dosing of additives and the injection of grout behind the segments.

In rock, large-diameter tunnels can be excavated by drilling and controlled blasting or or using tunnel boring machines which fragment the massif using high-speed high-resistance strength. The behaviour of the rock during excavation depends on the presence of fractures, zones of alteration and in-situ stress variations. For this reason, support designs must take into account the possibility of rockfalls, instability at the crown or differential movements in the side walls. The installation of primary and secondary supports must be carried out within strictly controlled timeframes to prevent the rock mass from losing cohesion or stiffness. Large-diameter tunnels also require a complex logistical design to enable the removal of excavated material, the transport of segments, the installation of electromechanical systems and the assurance of personnel safety in confined spaces.

CHAPTER 8 (Part 5): Tunnels, Underground Excavations and Ground Behaviour in Deep Excavations

8.17 Tunnels in hypersaturated environments and under extreme hydrostatic pressures

Tunnel construction in hypersaturated environments, such as areas near rivers, lakes, seas or deep aquifers, requires a thorough analysis of the extreme hydrostatic pressures acting on the excavation face and the structural lining. In these environments, water pressure can exceed several hundred kilopascals, exerting forces capable of destabilising the face if an appropriate construction method is not employed. The combination of high permeability, total saturation and steep hydraulic gradients can lead to phenomena such as upward flow, siphoning or internal erosion, all of which pose severe risks. To mitigate these effects, advanced techniques are used, such as face pressurisation in slurry tunnelling machines, pre-grouting to reduce permeability, frozen ground barriers and reinforcement using balanced pressure rings.

The design of tunnels under hypersaturated conditions requires linings capable of withstanding high external pressures throughout the project's service life. This involves taking into account material fatigue, hydrodynamic cycles related to fluctuations in the water table, and potential dynamic effects associated with tsunamis, storms or transient overloads. Furthermore, joint seals must be designed using special materials that ensure watertightness even under differential deformations of the lining. A crucial aspect is the management of infiltrating water, as even small leaks can cause progressive deterioration or trigger adverse chemical reactions. For this reason, internal drainage systems must be designed to capture and immediately evacuate any flow, preventing accumulations that increase internal pressure. Operation in hypersaturated environments requires continuous monitoring of hydraulic behaviour, allowing for adjustments to the machine's operating parameters and support systems.

8.18 Tunnels in karst regions and natural cavities

Karst regions present some of the most challenging conditions for tunnel construction, due to the presence of caverns, sinkholes, underground passages and voids formed by the dissolution of carbonate rocks.

These voids can range from small fissures to giant cavities that completely compromise the stability of the rock mass. In this type of terrain, the geotechnical properties are highly heterogeneous and depend on the degree of fracturing, the presence of clayey or sandy fill within the cavities, and the interaction with groundwater. Detecting voids requires advanced geophysical methods such as electrical tomography, ground-penetrating radar (GPR), refraction seismology or three-dimensional techniques based on surface waves. Without a thorough investigation, the risk of collapse during excavation increases significantly.

The construction strategy in karst terrain depends on the size and distribution of the cavities. Where the voids are small or medium-sized, consolidation grouting can be carried out to fill the cavities and reinforce the rock mass. When large cavities are detected, it may be necessary to alter the tunnel alignment or incorporate support structures designed specifically to pass through void zones. The use of tunnel boring machines in karst regions requires exceptional control of the face pressure, as a sudden loss of support can cause abrupt subsidence or flash flooding. The linings must be particularly robust and, in some cases, it is necessary to design sections reinforced with structural steel or high-strength concrete. Furthermore, cavities can interact with groundwater flows, increasing the risk of erosion or scouring, which requires specialised drainage systems and continuous monitoring of pore pressures.

8.19 Risk management in complex underground construction projects

Risk management in tunnelling is a comprehensive process that encompasses the identification, assessment, mitigation and monitoring of factors that could compromise the safety of the worksite, personnel and surrounding infrastructure. Risks may arise from unforeseen geological conditions, design errors, machinery malfunctions, fluctuations in the water table, seismic activity or unexpected variations in rock strength. To anticipate these events, geomechanical classification systems, probabilistic analysis, three-dimensional simulations and predictive models based on artificial intelligence are implemented. These approaches enable the evaluation of different scenarios and the definition of preventive measures before excavation begins. In many projects, alert levels are established which are triggered when the actual behaviour of the ground differs from predictions, making it possible to adjust the construction method in real time.

Risk management also involves contingency planning for extreme situations, such as sudden failures of the excavation face, flooding, partial collapses or failures of the primary support. Specific protocols are designed for each condition, including staff evacuation, emergency reinforcements and temporary changes to the advance sequence. In urban projects, risk management takes on an additional dimension, as settlement or vibrations can affect thousands of people and nearby critical infrastructure. Therefore, mitigation measures must be robust, redundant and based on real-time monitoring. Finally, comprehensive risk management does not end with construction; many underground works require ongoing monitoring to detect deferred deformations, changes in the water table and material degradation, ensuring their safety for decades to come.

8.20 Vibration and noise control during excavation

The vibrations generated during excavation can seriously affect sensitive buildings, laboratories, hospitals, heritage structures or high-precision equipment located near the site. The main sources of vibration include the operation of tunnel boring machines, the use of hydraulic hammers, controlled blasting, the movement of heavy machinery and ground compaction. To minimise these effects, strategies are implemented such as adjusting the tunnel boring machine's operating parameters, reducing the energy of explosives, finer fragmentation during drilling and blasting, the use of vibration-damping barriers, and planning appropriate schedules for highly vibratory activities. Vibration modelling allows the intensity of ground movement to be predicted at different points and permissible limits to be established in accordance with international standards and local conditions.

Noise is also a significant challenge, particularly in urban areas where night-time work can cause a nuisance to local residents.

To reduce noise levels, acoustic enclosures, sound-absorbing panels, low-noise equipment and operating techniques that minimise sudden impacts are used. Proper planning of internal transport routes and the reduction of unnecessary manoeuvres also help to minimise environmental noise. Some jurisdictions set strict noise and vibration limits, which require these variables to be continuously monitored. The data collected enables the modification of the excavation method, adjustment of the advance rate, or implementation of additional damping measures. Noise and vibration control is a crucial component in ensuring social acceptance of the project and protecting the integrity of nearby structures.

CHAPTER 8 (Part 6 – Final): Tunnels, Underground Excavations and Ground Behaviour in Deep Excavations

8.21 Design and operation of ventilation systems in deep tunnels

Ventilation in deep tunnels is a critical factor during both the construction phase and the subsequent operation of the infrastructure, due to the accumulation of gases, dust, vapours and pollutants generated by machinery or natural geological processes. During the construction phase, the ventilation system must be capable of removing gases from blasting, diesel engines and heavy machinery, as well as airborne particles that can harm workers' health and reduce visibility in the excavation. To achieve this, forced ventilation systems are used, comprising flexible ducts, high-power axial fans and pressurised ducts that drive clean air towards the working face and extract contaminated air via auxiliary galleries. The efficiency of these systems must be adjusted to the tunnel's advance, the volume of air required by the machinery and the total length of the excavation, as long tunnels experience pressure losses that must be compensated for by intermediate stations.

During the operational phase, the ventilation design depends on the tunnel's intended use, whether it be for vehicular traffic, railways, hydraulic systems or service tunnels. In road tunnels, for example, ventilation must control the concentration of nitrogen dioxide, carbon monoxide and exhaust particles, maintaining safe levels even during traffic congestion. Longitudinal systems use the airflow generated by jet fans, whilst transverse systems employ separate ducts for air intake and exhaust. In high-speed railway tunnels, the internal aerodynamics and the overpressure generated by the passing train are analysed, as this can induce sudden pressure variations and compromise passenger comfort. The operation of deep tunnels in densely populated areas requires continuous monitoring of air quality, temperature and the concentration of flammable gases, ensuring safe and efficient operation throughout their service life.

8.22 Tunnels for mass transit and specific geotechnical challenges

Tunnels used for mass transit systems, such as metros, commuter railways or multimodal transport hubs, present specific geotechnical challenges due to their interaction with existing underground networks, deep stations, ventilation shafts and sudden changes in geometry. In these systems, continuity of service is essential, so the design must ensure a high level of safety even under adverse conditions. The construction of deep stations involves large-volume excavations subjected to high lateral stresses, requiring robust retaining and shoring systems capable of withstanding significant deformations without affecting nearby structures. Interconnected tunnels must be aligned with centimetre precision, avoiding interference with building foundations, electrical cables or large-diameter pipes.

In densely built-up urban areas, the excavation of tunnels for mass transit must take into account interactions with existing buildings, sensitive structures, historic bridges and existing underground lines. The proximity of tunnels can induce stress interactions that alter the convergence of excavations or cause unexpected redistributions within the rock mass. To avoid these effects, advanced three-dimensional models are used to simulate the behaviour of multiple simultaneous excavations. Furthermore, vibrations generated by the frequent passage of trains must be controlled using isolation systems such as floating slabs, elastomeric bearings or anti-vibration technology integrated into the track. Continuous operation requires regular geotechnical maintenance, including crack inspection, convergence measurement, analysis of deferred settlements and verification of the hydrological behaviour of the surrounding environment.

8.23 Excavation using continuous mining and advanced mechanised systems

Continuous mining applied to civil engineering tunnels uses specialised machinery capable of excavating and supporting the ground without interruption, thereby increasing productivity and reducing the risks associated with traditional methods. Notable among these machines are roadheaders, which use cutting heads fitted with metal picks to break up the ground, and articulated arm cutters, used primarily in soils of medium hardness or soft rock. These technologies enable precise work in non-circular sections or complex geometries, such as service tunnels, underground caverns, underground stations or connecting tunnels. Continuous mining offers significant advantages in terms of safety, as it avoids the use of explosives and allows for more uniform control of excavation and support.

However, its use requires a detailed geotechnical analysis to ensure the stability of the working face and the immediate surroundings. The cutting capacity of the equipment can be affected by unexpected variations in rock hardness, the presence of rock inclusions, cavities or highly fractured zones. These conditions can cause the cutting arm to bounce or place excessive loads on the mechanical system, reducing its efficiency and generating excessive vibrations. To mitigate these risks, ground prediction systems based on torque sensors, acoustic analysis and vibration monitoring are employed, enabling cutting parameters to be adjusted before encountering adverse conditions. The combined use of continuous mining, immediate support and advanced monitoring ensures a superior level of control in complex excavations, particularly in urban environments or on projects where advance rate is a critical factor.

8.24 Underground caverns and large-scale excavations

Underground caverns are among the most impressive elements of underground engineering due to their size, geometry and structural complexity. Unlike conventional tunnels, which usually have relatively uniform cross-sections, caverns require massive excavations that can exceed 20 or 30 metres in clear span. These structures are used to house underground hydroelectric power stations, deep underground stations, water or fuel storage tanks, scientific facilities and emergency systems. Designing a cavern involves analysing the stability of the entire rock mass, taking into account the influence of natural stresses, fractures, discontinuities and ground heterogeneity. Excavation is carried out sequentially, using staged methods or controlled blasting that allow the e e equilibrium to be maintained during each phase.

The support of large-volume caverns requires complex systems comprising high-capacity bolts, anchored cables, steel arches, reinforced shotcrete and thick permanent linings. In some cases, prestressing systems are used to control initial deformation and generate a favourable stress state in the rock mass. The presence of water can increase pressure on the cavern walls or induce erosion in fractured zones, so extensive drainage systems are designed to prevent dangerous accumulations. Deep cavern excavations must also take into account long-term effects, such as rock creep, ground relaxation or chemical deterioration. Continuous monitoring using three-dimensional sensors allows the actual behaviour of the structure to be verified and the design to be adjusted as the excavation progresses.

8.25 Integration of digital technologies in underground construction projects

Digitalisation has profoundly transformed tunnel engineering through the use of technologies such as Building Information Modelling (BIM), real-time data analysis, smart sensors, predictive modelling and integrated monitoring systems. The use of BIM allows the entire project to be visualised in three dimensions, integrates geotechnical, structural, hydraulic and electromechanical information, and enables the construction sequence to be planned with an unprecedented level of precision. This methodology facilitates the early detection of conflicts, optimises the use of space and improves coordination between the various teams involved in the project. Furthermore, smart sensors located on the tunnel boring machine, the lining and in the ground provide continuous information on face pressure, advance, convergence, energy consumption, shield behaviour and environmental conditions.

Predictive systems based on artificial intelligence can analyse large volumes of data to identify patterns that predict failures or abnormal behaviour in the rock mass. This enables rapid, informed decision-making, allowing excavation parameters to be adjusted and significant risks to be avoided. The integration of advanced technologies in underground engineering improves the safety, efficiency and sustainability of the project, reducing costs and extending the tunnel's service life. In urban projects, digitalisation enables real-time communication of information to authorities, residents and operators, creating transparent and collaborative project management. The trend towards integrated digital systems will continue to grow in the coming decades, completely redefining the way tunnels are designed, built and operated.

CHAPTER 9

(Part 1): Stability of Slopes, Embankments and Earth Structures in Advanced Geotechnical Engineering

9.1 Introduction to the stability of slopes and earth structures

Slope stability is one of the cornerstones of modern geotechnical engineering, as it encompasses the study of the behaviour of soil and rock masses subjected to natural conditions and loads induced by human activities. Unlike other structural elements where materials behave in a relatively predictable manner, slopes exhibit complex behaviour due to soil variability, the presence of discontinuities, environmental changes and interaction with groundwater. Slope engineering ranges from natural slopes formed by geological processes to artificial slopes associated with roads, dams, railway embankments, deep excavations and industrial platforms. Each of these systems exhibits specific failure mechanisms and requires detailed analysis to ensure stability in both the short and long term.

The design of stable slopes requires an understanding of how internal stresses are distributed, how shear resistance is mobilised, and how external conditions influence the overall equilibrium. Furthermore, stability is not a permanent state; it can evolve over time due to factors such as wet-dry cycles, surface erosion, prolonged seepage, earthquakes, increases in load, and changes in land use. Consequently, geotechnical analysis must integrate concepts from hydraulics, soil mechanics, structural geology and geomorphology to prevent landslides, subsidence and mass movements. Errors in the design or monitoring of slopes can lead to catastrophic failures affecting roads, homes, bridges, dams and other critical infrastructure. For this reason, slope stability is considered an essential component of projects involving slope cutting, the creation of platforms or significant movement of natural earth.

9.2 Factors affecting slope stability

The stability of a slope depends on multiple interrelated factors that can alter the strength of the material, the geometry of the slope or the stresses acting upon it. Among the most influential factors are the geomechanical properties of the soil or rock, moisture content, topography, the depth of the water table, the presence of discontinuities, vegetation, human activity and the loads applied to the crest. Each of these factors can change over time, meaning that slopes require periodic assessments and continuous monitoring. For example, in clay soils, small variations in moisture content can significantly alter apparent cohesion and cause sudden changes in shear strength. In granular soils, the reduction in effective stress caused by a rise in the water table can trigger sudden failures, particularly during heavy storms or prolonged infiltration processes.

Other important factors relate to the internal structure of the terrain. The presence of stratification dipping towards the valley, open fractures, zones of alteration , or layers with marked differences in permeability can facilitate the formation of preferred fault surfaces. In rock, the interaction between joints, faults and planes of weakness determines the type of movement that could occur, such as planar slides, rotations, overturns or rockfalls. Human activity also plays a significant role; inadequate cutting, excessive loads at the top of the slope, deforestation, leaking pipes and poorly planned excavations can reduce the stability of a slope in a matter of weeks.

Furthermore, seismic events can generate accelerations that destabilise previously stable slopes, causing surface or deep landslides. Taken together, these factors make slope stability a complex phenomenon requiring detailed analysis and comprehensive solutions.

9.3 Classification of slopes and failure mechanisms

Slopes can be classified according to their origin, predominant material, geometry and potential failure depth. This classification makes it possible to identify the most likely mechanisms of instability and select appropriate mitigation measures. Natural slopes are formed by geological processes such as sedimentation, erosion, tectonic activity and weathering. In contrast, artificial slopes result from cutting and filling carried out during the construction of roads, mines, canals and industrial platforms. Within the material categories, slopes can be composed of soil, rock or mixed systems, where both materials interact to influence overall stability. Failure mechanisms vary widely and include rotational landslides, planar landslides, rockfalls, overturning, wedge movements, debris flows and progressive failures. Each type of movement is associated with particular geotechnical conditions that must be carefully assessed.

In cohesive soils, rotational landslides are common due to the material's tendency to mobilise resistance across curved surfaces. These landslides typically involve large volumes and may move slowly over long periods before experiencing rapid movement. In granular or stratified soils, planar landslides predominate, developing along weak planes or inclined interfaces. In rock, the geometry and orientation of discontinuities determine whether failure will occur through overturning, wedge sliding or individual rockfall. During heavy rainfall, debris flows or mudslides may occur, particularly on slopes with loose materials and low cohesion. Poorly compacted embankments and fill also may fail due to differential consolidation, internal erosion or loss of strength through saturation. Correct identification of the failure mechanism is essential for designing appropriate stabilisation solutions.

9.4 Methods for analysing slope stability

Slope stability analysis is based on assessing the balance between the forces tending to cause movement and those providing resistance. To this end, static, dynamic, deterministic and probabilistic analysis methods are used, depending on the complexity of the slope and the project requirements. Traditional methods include limit equilibrium theories, such as the Bishop, Janbu, Spencer and Morgenstern-Price methods, which allow the calculation of safety factors for various geometries and loading conditions . These methods are widely used due to their simplicity, ability to model complex failure surfaces and availability in commercial software. However, they must be supplemented with accurate information on soil properties and an adequate representation of the slope geometry.

Advanced numerical methods, such as finite element models, finite difference methods and discrete methods, enable the analysis of slopes under more realistic conditions, taking into account non-linear parameters, time-dependent deformations and changes in pore pressure. These models are useful for studying progressive failure, interaction between strata, the presence of reinforcement structures and seismic effects. They also enable the analysis of slopes during construction phases, the evaluation of the effectiveness of mitigation measures, and the monitoring of stability trends over time. Finally, probabilistic analyses incorporate the inherent variability of geotechnical parameters, assessing the probability of failure using Monte Carlo simulation or other statistical methods. This approach allows for the design of more robust solutions in projects where safety must account for significant ground uncertainties.

CHAPTER 9 (Part 2): Stability of Slopes, Embankments and Earth Structures

9.5 The influence of groundwater on slope stability

Groundwater is one of the most significant factors affecting slope stability, due to its ability to alter the effective strength of the soil, generate pore pressures and reduce the apparent cohesion of the materials. When the water table rises, effective stresses decrease, which reduces shear strength and can trigger movements even on slopes with relatively high safety factors. This phenomenon is particularly critical in saturated clay soils, where pore pressures rise rapidly during heavy storms or sustained seepage. The presence of water also promotes processes such as internal erosion, piping and the dissolution of fine particles, weakening the internal structure of the soil. Furthermore, water can infiltrate through discontinuities in fractured rock, causing hydraulic pressures that induce further fractures or rockfalls on rocky slopes.

Proper groundwater management is essential for maintaining stability throughout the life of the slope. To this end, surface drainage systems, lined ditches, deep horizontal drains, drainage galleries and relief wells are used to reduce pore pressure in critical areas. Deep horizontal drains are particularly effective, as they intercept water before it reaches the failure surface, gradually discharging the flow to safe points. On rock slopes, drainage boreholes are used to reduce pressure within fractures and cavities, preventing their progressive opening. Infiltration must also be controlled from the surface using vegetation cover, geomembranes and impermeable barriers. The interaction between water and soil affects not only immediate stability but also long-term behaviour, as prolonged cycles of wet and dry conditions can degrade the material's strength and lead to delayed failure.

9.6 Surface erosion and slope stability

Surface erosion is one of the most common processes of degradation on slopes and can lead to progressive instability if not properly controlled. This phenomenon occurs when rain, wind or surface runoff gradually removes soil particles, reducing the soil's cohesion and forming channels that encourage the concentration of flow. On steep slopes, erosion can rapidly develop into deep gullies that compromise overall stability. Sandy or silty soils are particularly vulnerable, as their particles break away easily and can be transported even by low-intensity flows. Furthermore, surface erosion can expose less resistant underlying layers, further accelerating the deterioration of the slope.

To mitigate these effects, various surface protection strategies are implemented, such as revegetation, the laying of geotextiles, hydroseeding, the installation of erosion control netting, and the construction of terraces that reduce water velocity. Vegetation contributes significantly to stability, as roots reinforce the soil and increase its apparent cohesion, whilst leaves and stems intercept the direct impact of rainfall. In areas where vegetation cannot establish itself immediately, biodegradable covers or geocells are used to stabilise the surface whilst the root system develops. On rocky slopes, wire mesh, surface anchors and rockfall protection systems are used to prevent landslides that could affect nearby infrastructure. Surface erosion must be monitored regularly, as small cracks or channels can quickly turn into major problems if not addressed in a timely manner.

9.7 Effects of seismic activity on slope stability

Seismic events represent one of the most serious threats to slope stability, due to the accelerations they induce in the ground and the temporary loss of strength they cause in saturated or loose materials. During an earthquake, seismic waves trigger rapid deformations that exceed the soil's capacity to respond, reducing shear strength and increasing the likelihood of shallow and deep landslides. The effect is particularly critical on slopes formed by non-cohesive soils, where liquefaction can cause an immediate loss of strength, transforming the ground into a fluid material that moves very rapidly. In fractured rock, vibrations can activate pre-existing planes of weakness, causing planar slides or rockfalls.

Seismic analysis of slopes requires an assessment of the expected maximum acceleration, the dynamic response of the ground, and susceptibility to liquefaction or softening. To this end, pseudo-static methods are used to introduce equivalent horizontal forces, alongside advanced dynamic models that simulate wave propagation through the rock mass. These models enable the identification of stress concentration zones, cumulative displacements and failure scenarios under different seismic intensities. On critical slopes, mitigation measures such as drainage systems to reduce pore pressures, retaining structures, geosynthetic reinforcements, anchored walls and deep foundations are applied to stabilise the areas at greatest risk. Post-earthquake monitoring is also essential, as seemingly small movements can develop into progressive failures weeks or months after the event.

9.8 Rock slopes and advanced stabilisation systems

Rock slopes exhibit behaviour that is highly dependent on fracturing, the orientation of discontinuities and the degree of alteration of the rock mass. Unlike soil slopes, where strength is determined by continuous parameters, rock slopes behave as discontinuous systems where blocks may slide, rotate, overturn or fall depending on the interaction between planes of weakness. Rock stability analysis requires identifying the geometry of the discontinuities through detailed geological mapping, stereographic analysis and geomechanical testing. Once the failure mechanisms have been identified, specific solutions are designed, such as passive or active anchors, friction bolts, prestressed cables, shotcrete and metal retaining mesh. These systems aim to bind unstable blocks together, increase apparent cohesion and reduce differential movements.

The stabilisation design must take into account environmental loads, the presence of water in fractures, thermal changes and potential seismic activity. On high slopes, dynamic barriers capable of absorbing the energy of falling blocks are used, preventing damage to infrastructure or roads located below. Protection systems such as high-strength mesh, suspended nets and geosynthetic-reinforced retaining walls enable the control of both small rockfalls and large-scale movements. Regular maintenance of rock slopes is essential, as temperature cycles, weathering and seepage can open up new planes of weakness or destabilise previously secure blocks. Early intervention and monitoring using drones, laser scanning or photogrammetry enable the detection of subtle changes in the rock mass's geometry that could lead to future failures.

CHAPTER 9 (Part 3): Stability of Slopes, Embankments and Earth Structures

9.9 Design and construction of embankments in geotechnical engineering

Embankments play a fundamental role in the construction of roads, railways, airports, industrial sites and infrastructure projects that require alterations to the natural topography of the land. Their design requires a thorough understanding of the behaviour of compacted soil, the interaction between layers, the distribution of stresses and the development of long-term settlement. Unlike natural slopes, embankments are artificially constructed structures, so their stability depends directly on the quality of the materials used, the degree of compaction, moisture control and the construction sequence. A poorly constructed embankment may experience differential settlement, landslides or excessive deformation that affect the functionality of the infrastructure. Therefore, the process of material selection, layering and compaction control is crucial to ensuring its performance.

The geotechnical design must take into account the material's shear strength, its maximum density, its permeability and its ability to dissipate pore pressures during and after construction. Furthermore, it is necessary to assess the slope gradient, the presence of water, seismic behaviour and possible internal failure mechanisms, such as the formation of critical surfaces within the embankment. In areas with variable climates, the behaviour of the embankment may change with cycles of humidity and temperature, which could lead to surface cracking or a loss of strength. For this reason, many projects incorporate reinforcement elements such as geogrids, geomesh or drainage layers that improve internal stability and reduce the risk of failure. The durability of an embankment also depends on controlling erosion on its surface, as slope degradation can lead to significant soil loss and affect overall stability.

9.10 Embankments on soft ground: challenges and solutions

Building embankments on soft ground presents one of the greatest geotechnical challenges due to the low strength and high compressibility of these materials. Soft soils, such as organic clays, highly plastic silts or peat, have a weak structure that can suffer excessive settlement or global failure if subjected to rapid or intense loads. When an embankment is placed on soft soil, the increase in load induces pore pressures that require long periods of consolidation to dissipate. If the construction process is too rapid, pore pressures may exceed the effective strength, causing shear failure or widespread subsidence. Consequently, construction on soft ground requires careful design incorporating methods of pre-consolidation, accelerated drainage and ground reinforcement.

Among the most commonly used solutions are preloading with or without vertical drains, the installation of gravel columns, the use of short or floating piles, partial soil replacement, and the incorporation of geogrids to distribute loads. Preloading involves applying a temporary additional load to accelerate soil consolidation before constructing the final embankment. Vertical drains allow water to escape more quickly, reducing the consolidation time. Gravel columns improve the soil's strength and its ability to redistribute stresses, whilst short piles allow the embankment to 'float' over weak surface layers. The design of these solutions must take into account the magnitude of permissible settlements, the service life of the infrastructure, and the interaction between the embankment materials and the foundation soil.

9.11 The use of geosynthetics to reinforce slopes and embankments

Geosynthetics have revolutionised geotechnical engineering in recent decades due to their ability to reinforce soils, control erosion, improve drainage and increase the stability of earth structures. Among the most commonly used types are geogrids, geotextiles, geocells and geomembranes, each serving specific functions within the design. Geogrids provide tensile strength, allowing the soil to be confined and loads to be distributed evenly. This is particularly useful in embankments built on soft soils, where reinforcement with geogrids reduces differential settlement and increases load-bearing capacity. Geotextiles act as filters and separators, preventing the migration of fine particles that could degrade the soil's properties or create areas of internal weakness.

Geocells, on the other hand, are three-dimensional structures that confine soil within flexible cells, improving its mechanical behaviour and reducing lateral deformation. This system is commonly used on steep slopes, high embankments and areas subject to dynamic loading. Geomembranes are used to control water infiltration, preventing underground flow from causing internal erosion or increasing pore pressures in critical areas. Designing with geosynthetics requires a detailed analysis that considers soil-reinforcement interaction, material durability, resistance to chemical agents, exposure to UV radiation and the structure's service life. Correct installation is essential for the system's performance, as defects in tensioning, joining or placement can significantly reduce the effectiveness of the reinforcement.

9.12 Modern slope stabilisation techniques

Modern slope stabilisation techniques combine advanced geotechnical principles with innovative technologies that improve soil strength and control movement. Among the most commonly used techniques are retaining walls, passive and active anchors, retaining piles, soil nailing, jet grouting and deep drainage systems. Soil nailing, for example, involves installing perforated steel bars grouted with cement slurry within the slope to create an internal reinforcement effect that increases overall strength. This technique is ideal for steep slopes where a flexible solution is required that adapts to the geometry and characteristics of the ground. Active anchors are used on high or unstable slopes, providing additional forces to counteract destabilising stresses.

Jet grouting enables the creation of high-strength soil-cement columns that act as structural barriers within the slope, improving both its stiffness and its drainage capacity. Retaining piles are suitable in areas where large masses of soil need to be retained or where loads induced by nearby buildings require more robust support. Retaining walls, whether made of reinforced concrete, gabions, mechanically stabilised walls or reinforced soil walls ( ), provide versatile solutions adaptable to different configurations. The choice of technique depends on factors such as the height of the slope, the type of soil, environmental conditions, the level of risk, the available budget and the expected service life of the project. The design must incorporate analyses of stability, deformations, drainage, soil-structure interaction and potential seismic effects.

9.13 Slopes in open-cast mining and special considerations

Slopes used in open-pit mining present unique conditions that require a specialised geotechnical approach due to their large scale, geological variability and the constant interaction between excavation and stability. Unlike road or urban slopes, mining slopes can reach heights of hundreds of metres and extend for several kilometres, implying significantly higher levels of risk. The design of mining slopes must take into account the orientation of discontinuities, the presence of geological faults, the effects of blasting, internal drainage and the progressive alteration of the rock mass due to prolonged exposure. Stability is assessed in both the short and long term, as mining operations continuously alter the geometry of the slopes and can trigger successive failures that compromise the safety of personnel and machinery.

Failure mechanisms in mining slopes include wedge-shaped movements, large-scale rotational slides, overturning, rockfalls, and failures induced by natural seismic activity or blasting. To prevent these, intensive monitoring techniques are implemented, including interferometric radar, automated total stations, 3D laser systems, GNSS sensors and photogrammetric analysis using drones. These systems enable the detection of millimetre-scale movements that indicate the development of a progressive failure. Mitigation measures may include redesigning slope angles, installing high-capacity anchors, constructing intermediate berms, controlling blasting, or improving internal drainage within the rock mass. Modern mining relies heavily on three-dimensional numerical models that simulate the evolution of the slope as the pit expands, enabling the anticipation of critical scenarios and the making of strategic decisions that maximise safety and operational efficiency.

CHAPTER 9 (Part 4): Stability of Slopes, Embankments and Earth Structures

9.14 Geotechnical monitoring of slopes and earthworks

Geotechnical monitoring has become an essential component in ensuring the stability of slopes and embankments, particularly on projects where geotechnical risks may evolve over time. This process involves the continuous or periodic measurement of parameters such as displacements, pore water pressure, seismic accelerations, groundwater level and internal ground deformations. The aim is to detect trends indicating the development of a progressive failure or the deterioration of initial conditions. Equipment such as inclinometers, extensometers, piezometers, load cells, accelerometers and automated surveying stations enable high-precision data to be obtained. On high slopes or in urban areas, remote monitoring systems using interferometric radar, laser scanners and drone-based photogrammetry enable the detection of millimetre-scale changes across large areas.

Analysis of the data collected enables the activation of early warning systems that alert to potential loss of stability. These systems are essential on slopes prone to slow-moving landslides, where displacement increases gradually before a complete failure occurs. In mining areas, monitoring can even help anticipate large-scale collapses by analysing the acceleration of movements. Integrating data with numerical models allows trends to be interpreted and future scenarios simulated, which is fundamental to decision-making. The installation of the monitoring system must be carefully designed, taking into account the location of the instruments, their depth, the type of soil, environmental exposure and the frequency at which data should be recorded. A well-implemented monitoring system can make the difference between detecting a failure in advance or facing catastrophic consequences with no chance of reaction.

9.15 Numerical modelling of slopes and embankments

Numerical modelling has become an essential tool for advanced stability analysis, enabling the simulation of complex conditions that traditional methods cannot assess accurately. Models based on

Finite element, finite difference and particle methods enable the actual geometry of the ground to be modelled, its non-linear behaviour to be characterised, and its response to variable loads to be assessed. These models can incorporate properties such as time-dependent stiffness, dynamic parameters, variations in permeability, progressive degradation or thermal effects. Modelling also allows for the study of construction sequences, the evaluation of excavation phases and the simulation of progressive failures that occur when the material's maximum strength is mobilised.

One of the greatest benefits of numerical modelling is the ability to evaluate different scenarios without the need for costly physical testing or intervention on the slope. Models can incorporate real geotechnical data, laboratory test results and monitoring measurements to adjust parameters and obtain more realistic representations of the ground's behaviour. In embankments, models help predict long-term settlement, stress distribution, interaction between layers and response to earthquakes or cyclic loads. However, the model's ac r accuracy depends directly on the quality of the input data and the engineer's judgement, as incorrect modelling can produce misleading results. Therefore, models must be calibrated using real data and validated by comparing them with behaviour observed in the field.

9.16 Stability of slopes subject to internal erosion and pipe installation

Internal erosion, also known as piping, is a silent yet potentially devastating phenomenon that can compromise the stability of slopes, embankments and retaining structures. This process occurs when water flows through the soil with sufficient energy to carry away fine particles, creating underground channels that weaken the internal structure of the material. Over time, internal erosion can lead to cavities, subsidence or global failure, particularly in loose soils or materials with non-uniform gradations. Piping is particularly dangerous on slopes near bodies of water, rivers, reservoirs or hydraulic embankments where hydraulic gradients can rise suddenly during floods or water level fluctuations.

The design of internal erosion protection includes granular filters, filter geotextiles, horizontal drains, discharge berms and pressure relief systems. Filters must be carefully designed to allow water to pass through without permitting the migration of particles from the underlying soil, which requires compliance with specific particle size criteria. On slopes subject to high hydraulic gradients, chimney drains, French drains and drainage geocomposites are also used to reduce internal pressure and prevent piping. Early detection is difficult, but signs such as increasing turbidity in the drained water, minor subsidence or longitudinal cracks may indicate the onset of the process. Internal erosion has caused historical failures in earth dams and road embankments, underscoring the importance of designing robust drainage systems and carrying out regular inspections.

9.17 Slope stability in coastal and marine environments

Slopes located in coastal areas or in marine environments are subject to extremely variable conditions, including waves, tides, currents, storms, saltwater intrusion and coastal erosion. These factors continuously alter the geometry of the slope and its mechanical properties. Waves can remove surface material, causing progressive instability, whilst tidal action can repeatedly saturate and desaturate the ground, reducing its apparent cohesion. Salt intrusion can affect soil chemistry, altering its internal structure, particularly in sensitive clays. In underwater environments, slopes may fail due to sediment overload, deep currents, seismic activity or massive landslides caused by unbalanced pressure conditions.

The stability of marine slopes also depends on the behaviour of sediment under conditions of low permeability and high saturation. Vibrations generated by storms or seismic activity can induce liquefaction in subsea sand, triggering submarine landslides that may affect cables, pipelines or energy infrastructure. To mitigate these risks, techniques such as rock revetments, wave protection structures, marine retaining walls, geotextiles resistant to saline environments and drainage systems designed to manage sea-level fluctuations are used. Works in coastal areas require a geotechnical design that considers not only static stability, but also the interaction between oceanic processes and the ground.

9.18 Progressive failures on slopes and the progression of damage

Progressive failures occur when a slope undergoes a redistribution of stresses that leads to the continuous weakening of the material until a global failure surface is activated. These processes usually begin with small deformations that do not pose an immediate risk, but which gradually increase as additional resistance is mobilised in critical zones. Progressive failure is characteristic of soils with brittle behaviour or of slopes subjected to repetitive loads, moisture cycles, internal erosion or prolonged saturation. In some cases, the process can take months or even years before resulting in a sudden collapse. Therefore, detecting the initial signs is crucial for implementing preventive measures.

Numerical models enable the study of the progression of a landslide, identifying areas where the mobilised resistance exceeds the available resistance. These areas act as the starting points for the landslide. Geotechnical instrumentation also allows the detection of slow movements that precede the collapse. A common strategy for mitigating progressive failures involves reducing destabilising stresses through drainage, reducing loads on the crest, installing reinforcements or modifying the slope geometry. In cases where the failure is already underway, emergency measures are implemented, such as support berms, temporary anchors or confinement using flexible retaining walls.

Bibliographical References

Alonso Pollán, J. A. (2014). *Gravel columns (GC)* [Master's thesis, Polytechnic University of Madrid].

Cedeño, J. G. R. (2015). *Study of the liquefaction phenomenon in Chile following the 2010 Maule earthquake* [Bachelor's thesis, Catholic University of Santa María].

Das, B. M. (2007). *Principles of foundation engineering* (6th ed.). Cengage Learning.

Das, B. M. (2010). *Fundamentals of foundation engineering*. Cengage Learning.

European Committee for Standardization. (2004). *EN 1998-5 Eurocode 8: Design of structures for earthquake resistance - Part 5: Foundations, retaining structures and geotechnical aspects.*

Geostru. (21 September 2020). *Soil liquefaction: what it is and how to address it.* https://blog.geostru.eu/es/licuefaccion-de-suelos-que-es-y-como-intervenir/

Keller Cimentaciones, S. L. U. (n.d.). *Dynamic compaction.* https://www.keller.com.es/experiencia/tecnicas/compactacion-dinamica

Lambe, T. W., & Whitman, R. V. (1969). *Soil mechanics*. John Wiley & Sons.

Menard Latam. (n.d.). *Gravel columns (CG).* https://www.menardlatam.com/soil-expert-portfolio/columnas-de-grava/

SAI - Soluciones Ambientales Integrales. (n.d.). *Applications of geosynthetics in civil engineering.* https://blog.sai.la/es/aplicaciones-de-geosinteticos-en- ingenieria-civil

SAI - Integrated Environmental Solutions. (n.d.). *Geocells: surface protection and stabilisation.* https://sai.la/es/productos/geoceldas

SAI - Integrated Environmental Solutions. (n.d.). *Geosynthetics: types, functions and applications.* https://sai.la/es/geosinteticos

Seed, H. B., & Idriss, I. M. (1971). A simplified procedure for evaluating soil liquefaction potential. *Journal of the Soil Mechanics and Foundations Division, 97*(9), 1249–1273.

Terzaghi, K. (1943). *Theoretical soil mechanics*. John Wiley & Sons.

Terzaghi, K., Peck, R. B., & Mesri, G. (1996). *Soil mechanics in engineering practice* (3rd ed.). John Wiley & Sons.

National Autonomous University of Mexico. (17 January 2021). *Excessive water extraction causes cities in Mexico to sink.* https://www.dgcs.unam.mx/boletin/bdboletin/2021_041_.html

University of Washington - Department of Civil & Environmental Engineering. (n.d.). *Niigata earthquake, 1964 - Liquefaction effects.* https://depts.washington.edu/liquefy/html/quakes/niigata/niigata.html

Wikipedia contributors. (n.d.-a). *Soil (engineering)*. Wikipedia. https://es.wikipedia.org/wiki/Suelo_(engineering)

Wikipedia contributors. (n.d.-b). *Soil mechanics*. Wikipedia. https://es.wikipedia.org/wiki/Mec%C3%A1nica

Yepes, V. (20 January 2014). *Dynamic compaction and its control using dynamic penetration tests*. Víctor Yepes' Blog. https://victoryepes.blogs.upv.es/2014/01/20/compactacion-dinamica/

Yepes, V. (2 July 2019). *High-pressure grouting: Jet grouting*. The Blog of Víctor Yepes. https://victoryepes.blogs.upv.es/2019/07/02/jet-grouting/

MIX
Papier aus verantwortungsvollen Quellen
Paper from responsible sources
FSC® C105338

Printed by Books on Demand GmbH, Norderstedt / Germany